Beginning Mathematica and Wolfram for Data Science

Applications in Data Analysis, Machine Learning, and Neural Networks

Jalil Villalobos Alva

Apress®

Beginning Mathematica and Wolfram for Data Science: Applications in Data Analysis, Machine Learning, and Neural Networks

Jalil Villalobos Alva
Mexico City, Mexico

ISBN-13 (pbk): 978-1-4842-6593-2
https://doi.org/10.1007/978-1-4842-6594-9

ISBN-13 (electronic): 978-1-4842-6594-9

Managing Director, Apress Media LLC: Welmoed Spahr
Acquisitions Editor: Steve Anglin
Development Editor: Matthew Moodie
Coordinating Editor: Mark Powers

Cover designed by eStudioCalamar

Cover image by Clint Adair on Unsplash (www.unsplash.com)

Distributed to the book trade worldwide by Apress Media, LLC, 1 New York Plaza, New York, NY 10004, U.S.A. Phone 1-800-SPRINGER, fax (201) 348-4505, e-mail orders-ny@springer-sbm.com, or visit www.springeronline.com. Apress Media, LLC is a California LLC and the sole member (owner) is Springer Science + Business Media Finance Inc (SSBM Finance Inc). SSBM Finance Inc is a **Delaware** corporation.

For information on translations, please e-mail booktranslations@springernature.com; for reprint, paperback, or audio rights, please e-mail bookpermissions@springernature.com.

Apress titles may be purchased in bulk for academic, corporate, or promotional use. eBook versions and licenses are also available for most titles. For more information, reference our Print and eBook Bulk Sales web page at http://www.apress.com/bulk-sales.

Any source code or other supplementary material referenced by the author in this book is available to readers on GitHub via the book's product page, located at www.apress.com/9781484265932. For more detailed information, please visit http://www.apress.com/source-code.

Printed on acid-free paper

To my family that supported me on all aspects.

Table of Contents

About the Author

Jalil Villalobos Alva is a Wolfram language programmer and Mathematica user. He graduated with a degree in engineering physics from the Universidad Iberoamericana in Mexico City. His research background comprises quantum physics, bioinformatics, and protein design. He has made contributions to AI in protein sciences, machine learning, and independent work on research involving topics of technology, blockchain, stochastic processes, and space engineering.

He also has a certification in Wolfram Language and participated in the Wolfram Summer School program, where he espoused and learned all the ideas behind the Wolfram Language.

His academic interests are focused on seeking the completion of a Master engineer Science in Engineering Studies in order to seek new challenges. During his free hours he likes to play soccer, swim, and listen to music.

About the Technical Reviewer

 **Mezgani Ali** is a Ph.D. student in Transmissions, Telecommunications, and Artificial Intelligence (National Institut of Postes and Telecommunications in Rabat) and researcher at Native LABs, Inc. He likes technology, reading and his little daughter Ghita. His first program was a Horoscope in Basic in 1993. He has done a lot of work on the infrastructure side of system engineering and software engineering and has managed networks and security.

He has worked for NIC France, Capgemini, and HP; he was part of the site reliability engineer's (SRE) team responsible for keeping data centers' servers and customers' applications up and running. He is fanatical about Kubernetes, REST API, MySQL, and Scala and is creator of the functional and imperative programming language PASP.

He holds a master's degree of ISTV Mathematics and Computer Science in France, and is preparing for classes for an engineering MPSI Degree in Morocco (His Youssef High School, Rabat).

Often he is a serious person, modest and relaxed, and is always looking for ways to constantly learn new things to improve himself.

Acknowledgments

I would like to thank Steve Anglin for the first approach and recommendations on starting up this project. I would also like to thank Mark Powers for establishing an effective communication link between all collaborators of this project. I also like to thank Mezgani Ali and Matthew Moodie for many useful comments and feedback during creation of this manuscript. They both helped me in improving the presentation of the material and theoretical work. And finally, I would like to thank my colleague Sebastian Villalobos, who helped a lot in the development of ideas, and Las Des Nestor and "Los Betos" for teaching me great mastery.

Introduction

Welcome to Beginning Mathematica and Wolfram for Data Science.

Why is data science important nowadays? Data science is an active topic that is evolving every day; new methods, new techniques, and new data is created every day. Data science is an interdisciplinary field that involves scientific methods, algorithms, and systematic procedures to extract data sets and thus have a better understanding of the data in its different structures. It is a continuation of some theoretical fields of data analysis such as statistics, data mining, machine learning, and pattern analysis. With a unique objective, data science extracts quantitative and qualitative information of value from the data that is being recollected from various sources, thus enabling one to objectively count an event, either for decision making, product development, pattern detection, or identification of new business areas.

Data Science Roadmap

Data science carries out a series of processes to solve a problem, including data acquisition, data processing, model construction, communication of results, and data monitoring or model improvement. The first step is to formalize an objective in the investigation. From the object of the investigation, we can proceed to the sources of the acquisition of our data. This step focuses on finding the right data sources. The product of this path is usually raw data, which must be processed before it can be handled. Data processing includes transforming the data from a raw form to a state in which it can be reproduced for the construction of a mathematical model. The construction of the model is a stage that is intended to obtain the information by making predictions in accordance with the conditions that were established in the early stages. Here the appropriate techniques and tools are used that are comprised of different disciplines. The objective is to obtain a model that provides the best results. The next step is to present the outcome of the study, which consists of reporting the results obtained and whether they are congruent with the established research objective. Finally, data monitoring has the intention to keep the data updated, because data can change constantly and in different ways.

Data Science Techniques

Data science includes analysis techniques from different disciplines such as mathematics, statistics, computer science, and numerical analysis, among others. Here are some disciplines and techniques used.

- Statistics: linear, multiple regressions, least squares method, hypothesis testing, analysis of variance (ANOVA), cross-validation, resampling methods

- Graph Theory: network analysis, social network analysis

- Artificial intelligence

- Machine learning

- Supervised learning: natural language processing, decision trees, naive bayes, nearest neighbors, support vector machine

- Unsupervised learning: cluster analysis, anomaly detection, K-means cluster

- Deep learning: artificial neural networks, deep neural networks

- Stochastic processes: Monte Carlo methods, Markov chains, time series analysis, nonlinear models

Many techniques exist, and this list only shows a part of them. Since research on data science, machine learning, and artificial neural networks are constantly increasing.

Prerequisites for the Book

This book is intended for readers who want to learn about Mathematica/Wolfram language and implement it on data science; focused on the basic principles of data science as well as programmers outside of software development—that is, people who write code for their academic and research projects, including students, researchers, teachers, and many others. The general audience is not expected to be familiar with Wolfram language or with the front-end program Mathematica, but little or any experience is welcome. Previous knowledge of the syntax would be an advantage to understand how the commands work in Mathematica. If this is not the case, the book provides the basic concepts of the Wolfram language syntax, the fundamental structure of expressions in the Wolfram language, and the basic handling and understanding of Mathematica notebooks.

Prior knowledge or some experience with programming, mathematical concepts such as trigonometric function, and basic statistics are useful; some understanding of mathematical modeling is also helpful but not compulsory.

Wolfram language is different from many other languages but very intuitive and easy to learn.

The book aims to teach the general structure of the Wolfram language, data structures, and objects, as well as rules for writing efficient code while also teaching data management techniques that allow readers to solve problems in a simple and effective way. We provide the reader with the basic tools of the Wolfram language, such as creating structured data, to support the construction of future practical projects.

All the programming was carried out on a computer with Windows 10 environment using version 12 of Wolfram Mathematica. Currently, Wolfram Mathematica also is supported in other environments such as Linux or macOS. The code found in the book will work with both the Pro and Student versions.

Conventions of the Book

Throughout the book, you may come across different words written distinctly from others. Throughout the book the words command, built-in functions, and functions may be used as synonyms that mean Wolfram language commands written in Mathematica. So a function will be written in the form of the real name (e.g., RandomInteger).

Evaluation of expression will appear with the Mathematica In / Out format, the same applies for blocks of code:

```
In[#]:= "Hello World"
Out[#]= "Hello World"
```

Book Layout

The book is written in a compact and focused way in order to cover the basic ideas behind the Wolfram language and to cover also details about more complex topics.

Chapter 1 discusses the starting topics of the Wolfram Language, basic syntax, and basic concepts with some example application areas, followed by an overview of the basic operations, and concludes by discussing security measures within a Mathematica session.

Chapter 2 provides the key concepts and commands for data manipulation, sampling, types of objects, and some concepts of linear algebra. It also covers the introduction to lists, which is an important concept to understand in the Wolfram language.

Chapter 3 discusses how to work properly with data and the initiation of the core structures for creating a dataset object, introducing concepts like associations and association rules. It concludes with remarks about how associations and dataset constructions can be interpreted as a generalization of a hash table aiming to expose a better understanding of internal structures inside the Wolfram language, with an overview of performing operations to a list and between lists. It then continues to discuss various techniques applied to dataset objects.

Chapter 4 exposes the main ideas behind importing and exporting data with examples throughout the chapter with various file formats. It also presents a very powerful command known as SemanticImport, which can import elements of data that are natural language.

Chapter 5 covers the topic areas for data visualization, common data plots, data colors, data markers, and how to customize a plot. Basic commands for 2D plots and 3D plots are also presented.

Chapter 6 introduces the statistical data analysis, starting with random data generation by introducing some common statistical measures followed by a discussion on creating statistical charts and performing an ordinary least square method.

Chapter 7 expose the basis for data exploration. It reviews a central discussion on the Wolfram Data Repository. This is followed by performing descriptive statistics and data visualization inside dataset objects, in Fisher´s Irises dataset.

Chapter 8 starts with concepts and techniques of machine learning, such as gradient descent, linear regression, logistic regression, and cluster analysis. It includes examples from various datasets like the Boston and Titanic datasets.

Chapter 9 introduces the key ideas and the basic theory to understand the construction of neural networks in the Wolfram language, such as layers, containers, and graphs, discussing the MXNet framework in the Wolfram language scheme.

Chapter 10 concludes the book by discussing the training of neural networks in the Wolfram language. In addition, the Wolfram Neural Net Repository is presented with an example application, examining how to access data inside Mathematica and the retrieval of information, such as credit risk modeling and fraud detection. The chapter concludes with the example of the LeNet neural network, reviewing the idea behind this neural network and exposing the main points on the architecture with the help of the MXNet graph operations and a final road map on the creation, evaluation, and deployment of predictive models with the Wolfram language.

CHAPTER 1

Introduction to Mathematica

The chapter begins with a preliminary introduction to why Mathematica is a useful and practical tool. So, core concepts of the Wolfram Language as well as its syntax will be examined and revised. The chapter will explain the internal structure of Mathematica, how it is designed, and how to insert code. The concept of a notebook will be introduced, which is important to understand the type of format that Mathematica handles. At this point we will see how a notebook can include code and text at the same time, considering that a notebook can be like a computable text file. Naturally, we will look at the parts or attachments that can be used in a notebook to help the user to better exploit the capabilities of code. The next section shows formally how to write expressions in Mathematica, examining topics such as arithmetic, algebra, symbols, global and local variables, built-in functions, date and time formats, simple plots of functions, logical operators, performance measures, delayed expressions, and accessing Wolfram alpha. We will then look at how Mathematica performs code computation, distinguishing between how to enter code in different forms of input, showing how to see what Mathematica interprets each time an evaluation is made. The chapter concludes with how to look for assistance within Mathematica, how to manage and handle errors as well as the search for a solution, and how to handle security in notebooks that incorporate dynamic content and how to handle them safely.

© Jalil Villalobos Alva 2021

J. Villalobos Alva, *Beginning Mathematica and Wolfram for Data Science*,

https://doi.org/10.1007/978-1-4842-6594-9_1

Why Mathematica?

Mathematica is a mathematical software package created by Stephen Wolfram more than 32 years ago. Its first official version emerged in 1988 and was created as an algebraic computational system capable of handling symbolic computations. However, over time Mathematica has established itself as a tool capable of performing complex tasks efficiently, automatically, and intuitively. Mathematica is widely used in many disciplines like engineering, optics, physics, graph theory, financial engineering, game development, software development, and others. Why Mathematica? Mathematica can be used where data management and mathematical computations are needed. So, Mathematica provides a complete integrated platform to import, analyze, and visualize data. Mathematica does not require plug-ins since it contains specialized functionalities combined with the rest of the system. It also has a mixed syntax given by its symbolic and numerical calculations. It allows us to carry out various processes without superfluous lines of code. It provides an accessible way to read the code with the implementation of notebooks as a standard format, which also serves to create detailed reports of the processes carried out. Mathematica can be described as a powerful platform that allows you to work effectively and concisely. Within computer languages, the Wolfram Language belongs to the group of programming languages that can be classified as a high-level multi-paradigm interpreted language. Unlike other programming languages, the Wolfram Language has unique rules in order to be able to write code in an understandable and compact way.

The Wolfram Language

Mathematica is powered by the Wolfram Language. Wolfram Language is an interpreted high-level symbolic and numeric programming language. To understand the Wolfram Language, it is necessary to keep in mind that the essential aspect of the language resembles a normal mathematical text as opposed to other programming languages syntax. The remarkable features of the Wolfram Language are:

- The first letter of a built-in function word is written in capital letter(s).

- Any element introduced in the language is taken as an expression.

- Expressions take values consisting of the Wolfram Language atomic expressions:

- – a symbol made up of letters, numbers, or alphanumeric contents

  – four types of numbers: integers, rational, real, and complex

  – the default character string is written within the quotation marks (" ")

- In Mathematica, there are three ways to group expressions.

  – Parentheses are used to group terms within an expression (expr1+expr2) + (expr3).

  – Commands entries are enclosed by brackets [ ]. Also, square brackets are used to enclose the arguments of a built-in function, F[x].

  – Lists are represented by the curly braces {}, {a, b, c}.

Structure of Mathematica

Before getting started with typing code we need to get the layout of Mathematica. To launch Mathematica, go to your Applications folder and select the Mathematica icon. The welcome screen should appear, as shown in Figure 1-1.

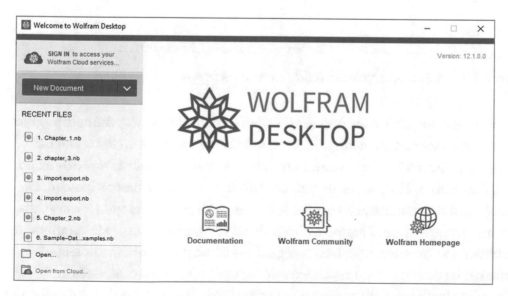

Figure 1-1. *This is the default's welcome screen for Mathematica version 12.1*

> **Tip**   A lot of useful information for first users and adept users can be found on the welcome page.

After the welcome screen appears, we can create a new document by selecting the button "New Document," and a blank page should appear like the one shown in Figure 1-2. New documents can be created by selecting File ➤ New ➤ Notebook or with the keyboard shortcut CTRL + N.

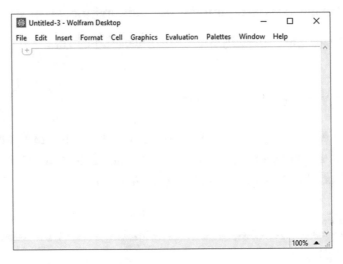

Figure 1-2. *A blank notebook ready to receive input*

The blank document that appears is called a notebook, and it's the core interaction between the user and Mathematica. Notebooks can be saved from the menu bar by selecting File ➤ Save, or Save as. Initializing Mathematica will always exhibit an untitled notebook. Notebooks are the standard format of a document. Notebooks can be customized so that they can expose text along with computations. However, the most important feature of Mathematica is that it can perform algebraic computations in addition to numerical calculations regardless of the purpose of the notebook. Mathematica notebooks are separated into input spaces called cells. Cells are represented by the square brackets on the right side of the notebook. Each input and output cell has its bracket. Brackets enclosed by large brackets mean that they are related computations, whether input or output. Grouped cells are represented by nested brackets that contain the whole evaluation cell. Other cells can also be grouped by selecting the cells and grouping them with the right-click option. Cells can also have the capability to show or hide input by simply double-clicking on the cells. To add a new cell, move the text cursor down, and a flat line should appear, marking the new cell ready to receive input expressions. The plus tab in the line is the assistant input tab, showing the various type of input supported by Mathematica. Figure 1-3 shows the input and output cells, which are grouped. The input cell is associated with In[-] and output cell with Out[-].

$$In[\bullet]:= \texttt{"Hello World"}$$

$$Out[\bullet]= \texttt{Hello World}$$

Figure 1-3. *Expression cells are grouped by input and output*

There are four main input types. The default input is the Wolfram Language code input. The Freeform is involved with Wolfram knowledge-base servers, and the results are shown in Wolfram Language syntax. Wolfram Alpha Query is associated with results explicitly shown as the Wolfram alpha website. And the external programming languages supported by Mathematica.

Tip   Keyboard shortcuts for front-end instruction commands are shown on the left side on each panel.

Design of Mathematica

Now that we have the lay of the land of Mathematica basic format, we can proceed to learn the internal structure of how Mathematica works. Inside Mathematica, there are two fundamental processes: the Mathematica kernel and the graphical interface. The Mathematica kernel is the one that takes care of performing the programming computations; it is where the Wolfram Language is interpreted and is associated with each Mathematica session. The Mathematica interface allows the user to interact with the Wolfram Language functions and at the same time document our progress. Each notebook contains cells, where the commands that the Mathematica kernel receives are written and then evaluated. Each cell has an associated number. There are two types of cells: the Input cell and the Output cell. These are associated with each other and have the following expressions: In[n]:= Expression and Out [n]: = Result or ("new expr"). The evaluations are listed according to which cell is evaluated first and continues in ascending order. In the case of quitting the kernel session, all the information, computations made, and stored variables will be relinquished, and the kernel will be restarted including the cell expressions. To quit a kernel session, select Evaluation on the toolbar followed by Quit kernel, then local.

Tip   To start a new kernel session, click Evaluation ➤ Start Kernel ➤Local.

To begin, try typing the computation:

```
In[1] := (11*17) + (4/2)
Out[1] = 189
```

The computation shows that In and Out have a number enclosed, and that number is the number associated to the evaluated expression.

As you might notice, a suggestion bar appears after every expression is evaluated (Figure 1-4) if you put the type cursor on the Output cell. Mathematica will always expose a suggestion bar, unless it is suppressed by the user, but the suggestion bar will show some suggestions of possible new commands or functions to be applied. The suggestion bar can sometimes be helpful if we are not sure what to code next; if used wisely, it might be of good assistance.

Figure 1-4. *Suggestion bar for more possible evaluations*

The input form of Mathematica is intuitive; to write in a Mathematica notebook you just have to put the cursor in a blank space, and the cursor will indicate that we are inside a cell that has not been evaluated. To evaluate a cell, click the keys [Shift + Enter], instructing Mathematica kernel to evaluate the expression we have written. To evaluate the whole notebook, go to the Evaluation tab on the toolbar, and select Evaluate Notebook. If the execution of calculations takes more time than expected, or you make a wrong execution of code, if you want to cease a computation, Mathematica provides several ways to stop calculations. To abort a computation, go to Evaluation ➤ Abort Evaluation, or for the keyboard shortcut, click [Alt + .].

When a new notebook is created, the default settings are assigned to every cell; that is the input style. Nevertheless, preferences can be edited in Mathematica with various options. To access them, go to Edit ➤ Preferences. A pop-up window appears (Figure 1-5) with multiple tabs (Interface, Appearance, Services, etc.). Basic customizations are included in the interface tab, such as magnification, language settings, and other general instructions. The Appearance Tab is related to code syntax color (i.e., symbols, strings, comments, errors, etc.). The Service Tab is related to the support contact team. Other tabs are related to more advanced capabilities that we will not use.

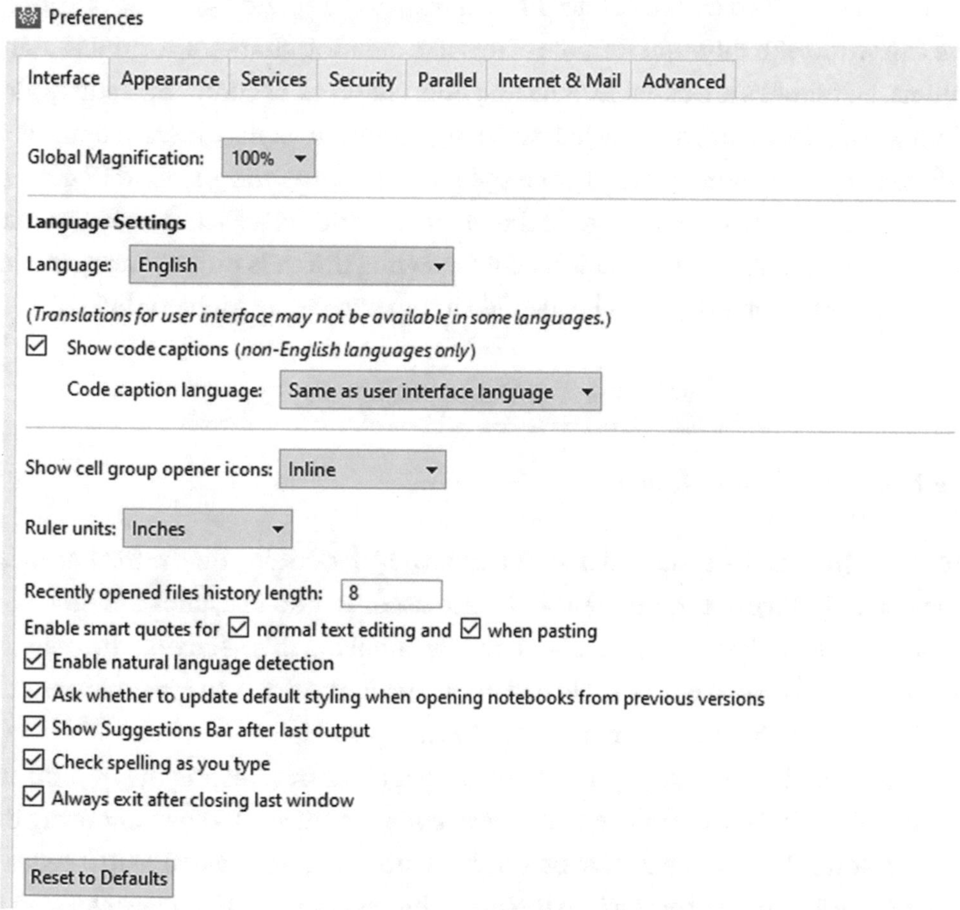

Figure 1-5. *Preferences window*

Notebooks

Text Processing

Notebooks can include explanatory text, titles, sections, and subsections. The Mathematica notebook more closely resembles a computable document rather than a programming command line. Text can be useful when a description of the code is needed. Mathematica allows us to input text into cells and create a text cell. Text cells can be text related to the computations we are doing. Mathematica has different forms to work with text cells. Text cells can have lines of text, and depending on the purpose of the text, we can work with different formats of text like creating chapters, formulas, items, and bullets. Notebooks are capable of having title, chapters, sections, and subsections just like a word-processing tool. By just selecting Format ➤ Style, different options will be exhibited. To have more control over a style cell, the formatting toolbar (Figure 1-6) can be used; to access the toolbar, go to the menu bar and select Window ➤ Toolbar Formatting. The formatting toolbar is useful for styling the cells rather than going into the menu bar every time. Text can be justified to the left, center, right, or full right.

Figure 1-6. *Style format toolbar*

The cell types can be arranged in different forms, depending on the format style a notebook uses. In Figure 1-7, different styles are used. To add text, click the Assistant tab input and select Plain text from the menu. By choosing this selection, the new cell created will be associated with simple text only. An alternative is to create a new cell and choose the text style from the formatting toolbox.

Styled text can be created with the formatting toolbar or selecting the desired style in Format ➤ Style ➤ "style" (title, chapter, text, code, input, etc.). When choosing the Style menu, you might notice the keyboard shortcuts for all the available text styles. This can be used instead of going into the menu bar every time. Plain text can also be converted into input text by formatting the cell in the Input style. There is no restriction in converting text; text can be converted into whatever style is supported in the format menu. To convert text, simply highlight the text or select the cell that contains the text. In Figure 1-7, multiple styles are used to create different results. Figure 1-7 shows the different styles that can be chosen in a notebook.

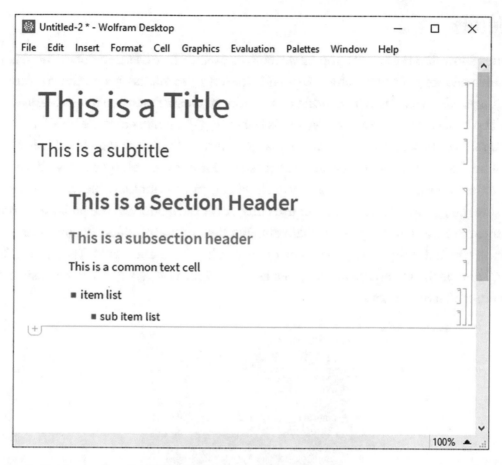

Figure 1-7. *A notebook with different format styles*

As shown in Figure 1-7, styled cells will look different from others. Each style has a unique order by which a notebook is organized into cell groups. A title cell has a higher order in a notebook, so the other cells are anchored to the title cell as shown in FIgure 1-7, but it does not mean that if another title cell is added both titles will be grouped. If the title cell is collapsed, the title will be the only displayed text.

Figure 1-7 shows multiple styles with their corresponding format; this includes, subtitle, section, subsections, plain text, item list, and subitem list.

Text can be given a particular style, changed, and different formats applied throughout the notebook. By selecting Font from the Format menu, a pop-up appears, this window allows changing the font, font style, size, and other characteristics.

Tip   To clear the format style of a cell, select the cell and then the right-click button and choose Clear Formatting.

Palettes

Palettes show different ways to enter various commands into Mathematica. There is a diverse quantity of special characters and typesetting symbols used in the Wolfram Language, which can be typed within expressions to closer resemble mathematical text. The best way to access these symbols is by using the pallets that are built in Mathematica. To select a simple pallet, just go to the toolbar, click Pallets -¿ Basic Math Assistant. Each pallet has different tabs that stand for different categories that have distinct commands as well as a variety of characters or placeholders that can be inserted by using the pallets. Hovering the mouse cursor over the symbol in the pallet will show the keyboard shortcut to type the character, function, or placeholder; to enter the symbol, type ESC followed by the name of the symbol, then ESC again. Try typing (ESC a ESC) to type the lowercase alpha Greek letter. In Figure 1-8, we can see the basic math assistant pallet of Mathematica.

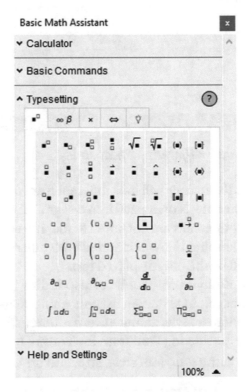

Figure 1-8. *This is the Basic Math Assistant palette*

Note   When putting the pointer cursor over a symbol or character, an info tip will pop up showing the keyboard shortcut.

Expression in Mathematica

Basic arithmetic operations can be performed on Mathematica, with a common intuitive form.

```
In[2] := (3*3) + (4/2)
Out[2] = 11
```

Mathematica also provides the capability to use a traditional mathematical notation. To insert a placeholder in the form, click [CTRL + 6]. To indicate the product operation, either use a space between expressions or add an asterisk (*)between.

```
In[3]:= 100²*10
Out[3]= 100000
```

```
In[4]:= 2*1
Out[4]= 2
```

The standard Mathematica format is to deliver the value closest to its regular form so that when dealing with decimal numbers or with general math notation, Mathematica always gives us the best precision; although it allows us to manipulate expressions numerically, to display numeric values, we use the function N[ ]. To insert the square root, type [CTRL + 2].

$$In[5]:= \frac{1}{2}+\sqrt{2}$$

$$Out[5]= \frac{1}{2}+\sqrt{2}$$

$$In[6]:=N\left[\frac{1}{2}+\sqrt{2}\right]$$

```
Out[6]= 1.91421
```

We can manage the number of precision of a numeric expression, in this case, we establish 10 decimal places.

```
In[7]:=N[77/12,10]
Out[7]= 5.923076923
```

For a shortcut to work with the decimal point, just type a dot (.) anywhere in the expression, and with this, you are telling Mathematica to calculate the value numerically.

$$In[8]:= \frac{4}{2}+\frac{2}{13}$$

```
Out[8]= 2.15385
```

Mathematica performs the sequence of operations from left to right, also according to how each expression is written, but it always follows the order of mathematical operations. To evaluate an expression without showing the result, we just add a semicolon (;) after the end of the first term. Like in the following example, the 11/7 is evaluated but not shown, and the other term is displayed.

```
In[9]:=11/7; Sqrt[4]
Out[10]= 2
```

The last form of code is called a compound expression. Expressions can be written in a single line of code, and with compound expression, expressions are evaluated in the corresponding sequence flow. If we write the semicolon in each expression, Mathematica will not return the values, but they will be evaluated.

```
In[11]:=3*4; 100*100; Sqrt[4];Power[2,2];
Out[11]=
```

As shown, there is no output but all of the expressions have been evaluated. Later we will use compound expressions to understand the concept of delayed expressions. This is a basic feature of the Wolfram Language that makes it possible for expressions to be evaluated but not displayed in order to consume memory.

Assigning Values

Each variable needs an identifier that distinguishes it from the others. A variable in the Wolfram Language can be a union of more than one letter and digit; it must also not coincide with protected words—that is, reserved words that refer to commands or built-in functions. (Keep in mind that the Wolfram Language is case-sensitive.) Mathematica allows assigning values to variables, also allowing the handling of algebraic variables.

Undefined variables or symbols will appear in blue font, as defined or recognized built-in functions will appear black. It is also true that the previously mentioned characteristics can be changed in the preferences window.

To write special constants and Greek letters, use the keyboard shortcut Esc pi Esc (pi number). A symbol of a vertical ellipsis (⋮) should appear every time ESC is typed. Another choice is witting the first letter of the name and a sub-menu showing a list of options should appear.

```
In[12]:= a=Pi
x=11
z+y
Out[12]= π
Out[13]= 11
Out[14]= y+z
```

In the previous example, Mathematica expresses each output with its cell, even though the input cell is just one. That is because Mathematica gives each cell a unique identifier. To access previous evaluations, the symbol (%) is used. Also Mathematica allows you to recover previous values using the cell input/output information by the command % # and the number of the cell or by writing explicitly the command with In [# of cell] or Out[# of the cell]. As shown in the next example, Mathematica gives the same value in each of the expressions.

```
In[9]:=
%14
In[14]
Out[14]

Out[15]= π
Out[16]= π
Out[17]= π
```

To find out whether a word is reserved within the Wolfram Language, use the Attributes command; this will display the attributes to the associated command. Attributes are general aspects that define functions in the Wolfram Language. When the word "Protected" appears in the attributes, it means that the word of the function is a reserved word. In the next example, we will see whether or not the word "Power" is reserved.

```
In[18]:= Attributes[Power]
Out[18]= {Listable,NumericFunction,OneIdentity,Protected}
```

As seen in the attributes, "Power" is a protected word. As you might notice with other attributes, one important attribute in the Wolfram Language is that most of the built-in functions in Mathematica are listable—that is, the function is interleaved to the lists that appear as arguments to the function.

Variables can be presented in a notebook in the following ways: (1) global variables, or those that are defined and can be used throughout the notebook, like the ones in the earlier examples; and (2) local variables, which are defined in only one block that corresponds to what is known as a module, in which they are only defined within a module. A module has the following form: Module [symbol1, symbol 2... body of module].

```
In[19]:= Module[{l=1,k=2,h=3}, h√(k+l) +k+l]]
Out[19]= 3+ 3√3
```

Variables inside a module turn green by default; this is a handy feature to know the code written is inside a module. A local variable only exists inside the module, so if you try to access them outside their module, the symbol will be unassigned, as shown in the following example.

```
In[20]:= {l,k,h}
Out[20]= {l,k,h}
```

Variables can be cleared with multiple commands, but the most suitable command is the Clear[symbol], which removes assigned values to the specified variable or variables. So, if we evaluate the variable after Clear, Mathematica would treat it as a symbol, and we can check it with the command Head; Head always gives us the head of the expression, which is the type of object in the Wolfram Language.

```
In[21]:= Clear[a,x,y]
```

And if we check the head a, we would get that "a" is a symbol.

```
In[22]:= Head[a]
Out[22]= Symbol
```

Symbols or variables assigned during a session will remain in the session memory unless they are removed or the kernel session is ended.

Note   Remove is an alternative to Clear.

Built-in Functions

Built-in commands or functions are written in common English and have the first letter
capitalized. Some functions have abbreviations, and others have two capital letters.
Here we present different examples of functions. Built-in functions are the way to group
expressions and statements so that they are executed when they are called; many receive
arguments. An argument is a value or values that the function expects to receive when
called, in order to execute the function correctly. A function may or may not receive
arguments; arguments are separated by a comma.

```
In[23]:= RandomInteger[]
Out[23]= 0
```

Note   RandomInteger, with no arguments, returns a random integer from the
interval of 0 to 1, so do not panic if the result is not the same.

```
In[24]:= Sin[90 Degree] + Cos[0 Degree]
Out[24]= 2

In[25]:= Power[2,2]
Out[25]= 4
```

Built-in functions can also be assigned symbols.

```
In[26]:= d=Power[2,2]
F=Sin[π] +Cos[π]
Out[26]= 4
Out[27]= -1

In[28]= Clear[d,f]
```

Some commands or built-in functions in Mathematica have available options that can be specified in a particular expression. To see whether a built-in function has available options, use Option. As in the next example, the function RandomReal is used for creating a pseudo-random real number between an established interval.

```
In[29]:= Options[RandomReal]
Out[29]= {WorkingPrecision → MachinePrecision}
```

As you may notice, RandomReal has only one option for specifying particular instructions inside the command. This is WorkingPrecision, and the option value as default is MachinePrecision. The WorkingPrecision option defines the number of digits of precision for internal computations, whereas MachinePrecision is the symbol used to approximate real numbers indicated by $MachinePrecision to see the value of MachinePrecision type $ MachinePrecision. In the next example, we will see the difference between using an option with the default values and using costume values.

```
In[30]:=RandomReal[{0,1}, WorkingPrecision → MachinePrecision]
RandomReal[{0,1}, WorkingPrecision → 30]
Out[30]= 0.426387  0.163331659063026438061773609723

Out[31]= 0.163331659063026438061773609723
```

Tip   Environmental variables always start with the dollar sign (e.g., $MachinePrecision).

As you may notice, the first one returns a value with 6 digits after the decimal point, and the other returns a value with 30 digits after the decimal point. But some other built-in functions do not have any options associated, like Power.

```
In[32]:=Options[Power]
Out[32]= {}
```

Dates and Time

The DateObject command provides results for manipulating dates and times in a concrete way. Date and time input and basic words are supported.

```
In[33]:=DateObject[]
Out[33]=
```

Figure 1-9. *The date of Wed 10 Jun 2020*

DateObjects with no arguments give the current date, as can be seen in Figure 1-9. Natural language is supported in Mathematica—for instance, getting the date after Wed 10 Jun 2020 (Figure 1-10).

```
In[34]:= Tomorrow
Out[34]=
```

Figure 1-10. *The date of Thu 11 Jun 2020*

The date format is entered as year, month, and day (Figure 1-11). It also supports different calendars, as shown in Figures 1-12 and 1-13

```
In[35]:= DateObject[{2020, 6, 10}]
Out[35]=
```

Figure 1-11. *The date of Wed 10 Jun 2020*

```
In[36]:=
DateObject[Today, CalendarType → "Julian"]
```

```
DateObject[Today, CalendarType → "Jewish"]
Out[36]=
```

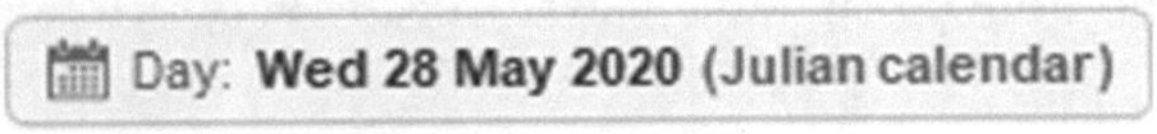

📅 Day: **Wed 28 May 2020** (Julian calendar)

Figure 1-12. *Today's date on the Julian calendar*

```
Out[37]=
```

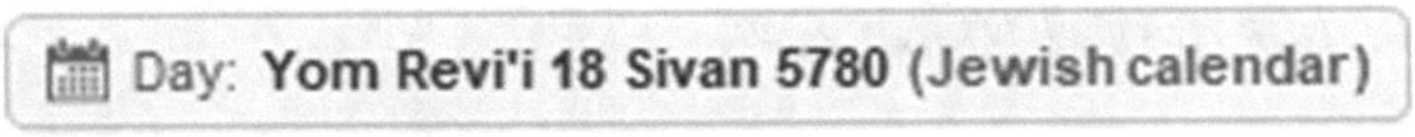

📅 Day: **Yom Revi'i 18 Sivan 5780** (Jewish calendar)

Figure 1-13. *Today's date on the Jewish calendar*

The command also supports options that are related to a time zone (Figure 1-14).

```
In[38]:= DateObject[{2010,3,4}, TimeZone → "America/Belize"]
Out[38]=
```

📅 Day: **Thu 4 Mar 2010**

Figure 1-14. *Belize time zone for the input date*

Sunset and sunrise of our current location can be calculated (Figure 1-15 and Figure 1-16).

```
In[39]:= Sunset[Here, Now]
Sunrise[Here, Yesterday]
Out[39]=
```

📅 Minute: **Mon 28 Sep 2020 19:27** GMT−5.

Figure 1-15. *Mon 28 Sep 2020, sunset time in the location of GMT-5*

```
Out[40]=
```

To get the current time, use TimeObject with zero arguments (Figure 1-17). It can be entered in the format of 24h or 12h. To introduce the time, enter the hour, minute, and second.

📅 Minute: **Sun 27 Sep 2020 19:28** GMT−5.

Figure 1-16. *Sun 27 Sep 2020, sunset time in the location of GMT-5*

```
In[41]:= TimeObject[]
Out[41]=
```

🕐 **23:47:50**

Figure 1-17. *Wed 10 Jun GMT-5 time*

Time zone conversion is supported; convert 5 p.m. from GMT-5 Cancun time to Pacific Time Los Angeles (Figure 1-18).

```
In[42]:= TimeZoneConvert[TimeObject[{17,0,0}, TimeZone → "America/Cancun"],
"America/Los Angeles"]
Out[42]=
```

🕐 **15:00:00** PDT

Figure 1-18. *Cancun 5 p.m. time conversion to Los Angeles PDT*

Basic Plotting

The Wolfram Language offers a basic description to easily create graphics. So we can create graphics in two dimensions and three dimensions. It also has a wide variety of graphics such as histograms, contour, density, and time series, among others. To graph a simple mathematical function we use the Plot command, followed by the symbol of the variable and the interval where we want to graph (Figure 1-19).

```
In[43]:= Plot[x³,{x,-20,20}]
Out[43]=
```

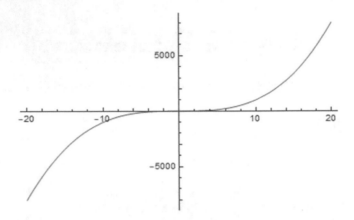

Figure 1-19. *A cubic plot*

The plot function also supports the handling of more than one function; simply gather the functions inside curly braces.

```
In[44]:= Plot[{Tan[x], x},{x,0,10}]
Out[44]=
```

Figure 1-20 shows the two functions in the same graph; each function has a unique color.

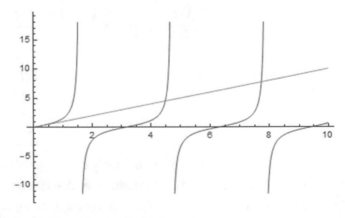

Figure 1-20. *Multiple functions plotted*

We can also customize our graphics in color if the curve is thick or dashed; this is done with the PlotStyle option (Figure 1-21).

```
In[45]:= Plot[{Tan[x], x},{x,0,10}]
Out[45]=
```

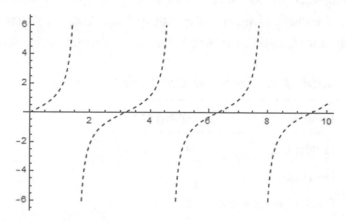

Figure 1-21. *Dashed tangent function*

The PlotLabel option allows us to add basic descriptions to our graphics by adding a title. On the other hand, the AxesLabel option lets us add names to our axes, both x and y, as depicted in Figure 1-22.

```
In[46]:= Plot [eˣ,{x,0,10},PlotStyle→{Blue},PlotLabel→"eˣ",AxesLabel→
{"x-axis", "y-axis"}]
Out[46]=
```

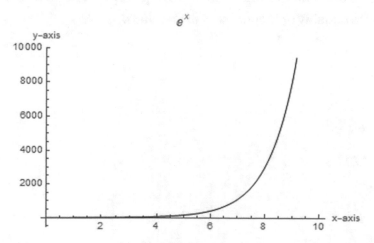

Figure 1-22. *A plot with title and labeled axes*

Logical Operators and Infix Notation

Infix notation and logical operators are used commonly in logical statements or comparison of expressions, and the result values are either true or false. Table 1-1 shows the relation operators of the Wolfram Language. Table 1-1 shows general options for 3D graphics.

Table 1-1. *Operators and Their Definitions*

Definition	Operator Form
Greater than	<
Less than	<
Greater than or equal	≥
Less than or equal	≤
Equal	=
Unequal	!= or ≠

Relational operators, also called comparison and logical binary operators, are used to check the veracity or falsity of certain relationship proposals. The expressions that contain them are called relational expressions. They accept various types of arguments and the result can be true or false—that is, they can be Boolean results. As we can see, they are all binary operators, of which two are of equality condition = = and !=. These serve to verify the equality or inequality of expressions.

```
In[47]:= 6*1 > 2
Out[47]= True

In[48]:= 6*1 < 2
Out[48]= False
```

$$In[49]:= \frac{1}{2} \geq 1/2$$
```
Out[49]= True
```

$$In[50]:= 1/4 \leq \frac{1}{2}$$
```
Out[50]= True

In[51]:= 3.12 == 2.72
Out[51]= False
```

```
In[52]:= π != √-1
Out[52]= True
```

Boolean operands are used to produce a true or false result or to test whether a condition is satisfied. Table 1-2 shows Boolean operators of the Wolfram Language. Boolean operators produce a Boolean result, which is used through expressions.

Table 1-2. *Boolean Operators and Their Definitions*

Definition	Operator Form
AND	&& or ∧
OR	‖ or ∧
XOR	⊻
Equivalent	⇔
Negation	¬

The AND operator returns a true value if both expressions are true. Otherwise the result is false.

```
In[53]:= 2==1 &&3.12==2
Out[53]= False
```

The OR operator returns true if any of the expressions is true. Otherwise it returns false. This operator has an analogous operation to the previous one.

```
In[54]:= 2*2==3||23*2==1
Out[54]= False
```

XOR operator is an exclusive or operator and returns true when both expressions are different. Otherwise it returns false when the expressions have the same value.

```
In[55]:= 2==1 ⊻ 2==2
Out[55]= True
```

The equivalent operator returns true if expressions are provable from each other. Otherwise it returns false.

```
In[56]:= Power[1,2] ⇔ 1²
Out[56]= True
```

The negation operator, also called logical negation, returns a value that can be an expression that evaluates to a result. The result of this operator is always a Boolean type.

```
In[57]:= \[Not]2==1
Out[57]= True
```

Another approach, instead of using Boolean operators, is using different functions with post fix (Q), which consists of testing whether an object meets the condition of the built-in function. A few honorable mentions are SameQ, UnsameQ, AtomQ, IntegerQ, and NumberQ. In the next example, we are going to test whether a number is a float expression or an integer.

```
In[58]:=
IntegerQ[1]
IntegerQ[1.]
Out[58]= True
Out[59]= False
```

A valuable application of a function called AtomQ can tell us whether an expression is subdivided into subexpressions. Later we will see how to deal with subexpressions with lists. If the result is true, then the expression cannot be subdivided into subterms, and if it is false, then the expression has subterms.

```
In[60]:= AtomQ[12]
Out[60]= True
```

As shown, numbers cannot be subdivided, because a number is a canonic expression, the same apllies for strings.

Algebraic Expressions

The Wolfram Language has the capability to work with algebraic expressions. For instance, perform symbolic computations, algebraic expansions, and simplifications, among others. You will notice that many words used in common language in algebra are preserved in Mathematica. To expand an algebraic expression use Expand.

```
In[61]:= Expand[((x^2)+y^2)*(x+y)]
Out[61]= x^3+x^2 y+x y^2+y^3
```

Adding a space between variables is the same as adding the multiplication operator. This can be checked by a*x==a x.

```
In[62]:= Expand[a x^2*(a x)^3]
Out[62]= a^4 x^5
```

But be careful when dealing with writing algebraic expression, because the symbol ax is not the same as a x. This also is checked using the SameQ[ax, a x] or the short notation a x === ax. To simplify an expanded expression, use Simplify or FullSimplify.

```
In[63]:= Simplify[x^3+x^2 y+x y^2+y^3]
FullSimplify[x^3+x^2 y+x y^2+y^3]
Out[63]= x^3+x^2 y+x y^2+y^3
Out[64]= (x+y) (x^2+y^2)
```

The difference is that the latter tries transformations to simplify the expression more broadly. To unite terms over a repeated denominator, use Together. To expand into partial fraction decomposition, use Apart.

```
In[65]:=
```

$$\text{Together}\left[\frac{1}{z}+\frac{1}{z+1}-\frac{1}{z+2}\right]$$

$$\text{Apart}\left[\frac{2+4z+z^2}{z(1+z)(2+z)}\right]$$

$$\text{Out[65]}=\frac{2+4z+z^2}{z(1+z)(2+z)}$$

$$\text{Out[66]}=\frac{1}{z}+\frac{1}{1+z}-\frac{1}{2+z}$$

Solving Algebraic Equations

Various functions are accessible for finding solutions to algebraic equations. The most common is Solve. The first argument is the equation or expression to be solved, and the second is for the variable to solve.

Note   As you might remember, equal is expressed as double equal (==); do not use one equal (=) because that means assigning a value to a symbol or variable.

```
In[67]:= Solve[z^2+1==2,z]
Out[67]= {{z→-1},{z→1}}
```

The result means that z has two solutions: one is -1 and the other 1. Each result is expressed in the form of a rule. A rule expression is used to change the assignment of the left side to the one on the right side (left → right) whenever it applies. For example, z → 1 is the same as Rule Rule[z,1].

To verify the solution, the values of z (-1,1) need to be replaced in the original equation. For this we can use the ReplaceAll operator (/.) along with the rule command → or Rule, which is used to apply a transformation to a variable or a pattern with other expressions.

```
In[68]:= z^2+1/.Rule[z, {1,-1}]
Out[68]= {2,2}
```

The other option is to type the solutions explicitly in the equation.

```
In[69]:= {1^2+1==2,(-1)^2+1==2}
Out[69]= {True,True}
```

Multiple equations can be solved too, given a system of equations and a list of interested variables. To solve the equations, place the system of equations in one list and the variables in another list.

Note   As you might notice, the results are given in the form of list. Lists are essential structures in the Wolfram Language and are discussed in the next chapter.

For example, solve the next system of equations.

$$x+y+z==2$$
$$6x-4y+5z==3$$
$$5x+2y+2z==1$$

The solution is

In[70]:= Solve[{x+y+z==2,6x-4y+5z==3,x+2y+2z==1},{x,y,z}]

Out[70]= $\left\{\left\{x \to 3, y \to \dfrac{10}{9}, z \to -\dfrac{19}{9}\right\}\right\}$

The latter process is also applicable for equations assigned to variables. We can write this with the use of compound expressions.

In[71]:= EQ1=x+y+z==2;EQ2=6 x-4 y+5 z==3;EQ3=x+2 y+2 z==1;
Solve[{EQ1,EQ2,EQ3},{x,y,z}]

Out[72]= $\left\{\left\{x \to 3, y \to \dfrac{10}{9}, z \to -\dfrac{19}{9}\right\}\right\}$

Solve also works with pure algebraic equations.

In[73]:= Solve[{x+y+z==a,6x-4y+5z==b,x+2y+2z==c},{x,y,z}]

Out[73]= $\left\{\left\{x \to 2a-c, y \to \dfrac{1}{9}(7a-b-c), z \to \dfrac{1}{9}(-16a+b+10c)\right\}\right\}$

Solve supports expressions with a mixture of logical operators, expressing y and x in terms of z.

In[74]:= Solve[EQ1&&EQ2,{x,y}]

Out[74]= $\left\{\left\{x \to \dfrac{1}{10}(11-9z), y \to \dfrac{9-z}{10}\right\}\right\}$,

using the OR operator.

In[75]:= Solve[x^2+y^2==0||x-2y==1,x]
Out[75]= {{x→-I y},{x→I y},{x→1+2 y}}

Solve returns the solution for each of the equations entered. Establishing a condition with the AND operator we can return solutions that satisfy a condition, for example the equation below has two solutions 1 and -1, but we can solve the equation with the condition that z must be different from 1.

In[76]:= Solve[z^-2+1==2&&z!=1,z]
Out[76]= {{z→-1}}

In order to obtain more general results, Reduce is used. For example,

In[77]:= Reduce[Cos[x]==-1,x]

Out[77]= $c_1 \in Z$ & & $(x = = -\pi + 2\pi c_1 \,\|\, x = = \pi + 2\pi c_1)$

Here the alternative solutions are separated by the OR operator, and the condition is established by the AND. So, this means that there are two possible solutions $-\pi + 2\pi c_1$ or $\pi + 2\pi c_1$, and that the constant c_1 must be a number that belongs to the integers (Z). In addition, Reduce can also solve inequalities.

In[78]:= Reduce $[h^2 + k^2 < 11, \{h, k\}]$

Out[78]= $-\sqrt{11} < h < \sqrt{11}$ & & $-\sqrt{11 - h^2} < k < \sqrt{11 - h^2}$

Here the simultaneous equations are for h and k. Furthermore, Reduce can show the combination of equations with certain conditions.

In[79]:= Reduce $[\alpha + \beta * \alpha \wedge 2 = = E, \alpha]$

Out[79]= $(\beta == 0 \,\&\&\, \alpha == e)$ $\left| \left(\beta \neq 0 \,\&\&\, \left(\alpha == \dfrac{-1 - \sqrt{1 + 4e\beta}}{2\beta} \quad \alpha == \dfrac{-1 + \sqrt{1 + 4e\beta}}{2\beta} \right) \right) \right.$

The first solution is that α and β must be the number e and zero. The second solution is in terms of α and the condition that β must be different from zero.

Using Wolfram Alpha Inside Mathematica

A really good application inside Mathematica is the use of the Wolfram alpha computable knowledge-base inside Wolfram Mathematica. Wolfram Alpha can be called from Mathematica with the Wolfram Alpha Query. To enter the Wolfram Alpha query, type the double equal sign before typing any expression; an orange asterisk with a white equal sign should appear, meaning that input typed will be a query with natural language. To execute the cell, just click the enter key.

So, for example, algebraic equations can be solved using the Wolfram Alpha query.

In[80]:=

Out[80]:=

Solve y^2 + x^2 == 1 && y + 2 x == 0, for {x, y}

Figure 1-23. *Wolfram Alpha query input*

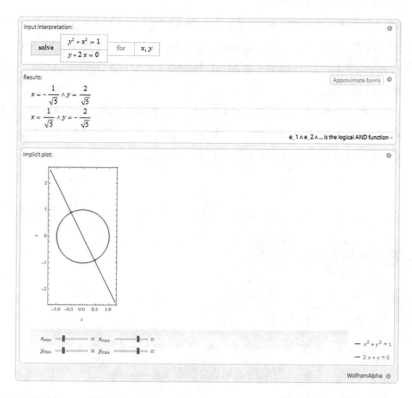

Figure 1-24. *Wolfram Alpha query output*

Figures 1-23 and 1-24 shows the input and the output of the Wolfram Alpha query.

As shown, the system returns the solutions for x and other calculations. The aforementioned cell represents the calculations in the Wolfram Alpha form, clicking the plus icon will show a list of different forms of input. To see equivalent in the Wolfram Language, select the Input option. The other related way to use Wolfram Alpha is with freeform input. It is worth mentioning that words associated with Mathematica commands, like Reduce, can be used too.

In[81]:=

```
Reduce x^2+y^2==1&&y+2x==0, for {x,y}
Reduce[{x^2 + y^2 == 1, y + 2*x == 0}, {x, y}]
```

Figure 1-25. *Input code in the freeform input*

$$\text{Out[81]:=} \left(x == -\frac{1}{\sqrt{5}} \quad x == \frac{1}{\sqrt{5}} \right) \&\& y == -2x$$

Figure 1-25 shows the input cell in the freeform input. Clicking the plus icon will show more calculations like in the Wolfram Alpha query. The following code is the equivalent in the Wolfram Language of input typed. Clicking the code will replace it the Wolfram Language syntax.

A clarification here, not just calculations can be made. With the Wolfram Alpha, access to curated data is available from various topics. For example, getting the financial data for a particular stock or the population of Australia.

```
In[82]:=
Out[82]:=
```

Tesla stock March	
⤷ Result	

mean	$111.82 (US dollars)
highest	$149.90 (US dollars) (Wednesday, March 4, 2020)
lowest	$72.24 (US dollars) (Wednesday, March 18, 2020)
volatility	146%
change	−30%

Figure 1-26. *Input and output of the summary of the Tesla stock on March 2020*

Figure 1-26 shows a summary of how the stock performed in the month of March.

```
In[83]:=
```

Figure 1-27. *Input for the population of Australia*

```
Out[83]:= 25 203 200 people
```

And Figure 1-27 shows the population of Australia and also displays the code of the Wolfram Language.

Both freeform input and Wolfram Alpha queries can be very useful and practical tools. For example, in the sense that you do not know the appropriate syntax of a function or command, try using the freeform input in natural language so that when evaluated, you can get the equivalent Wolfram Language syntax of that function. Nevertheless, a downside of the Wolfram Alpha query is that the computations are done outside Mathematica, meaning that the computations are made in the Wolfram Alpha servers, whereas calculations with freeform input can be reproduced inside Mathematica. In other words, sometimes it is preferable to work directly with the Wolfram Language in order to have a better management of the results, as it can be tedious to extract results from Wolfram Alpha. It should be noted that to access these two features from Mathematica, it is necessary to have access to Wolfram servers with an online network.

Delayed Expressions

The Wolfram Language has two important features. First, let's explain how the Set works. The symbol = is the script for Set and := for SetDelayed. The Set mechanism is represented by W ="expr"; it means that Mathematica evaluates the expression straightaway, then each time the variable or defined function is called, the value of W is written and the result is shown. On the contrary, using W := "expr" it stands that the expression will not be evaluated until called, so for each time the "W" is called, it will evaluate a new expression every time.

```
In[84]:= W=RandomReal[]
Out[84]= 0.552303
```

We test if W equals W.

```
In[85]:= W==W
Out[85]= True
```

In this case, the condition is true because we used Set for declaring the variable W with the function RandomReal, which returns a pseudo-random choice from 0 to 1. The same approach will be used but for SetDelayed, and the result will be false, because every time W appears, the function will be called to a new evaluation. We can write the code in a compound expression.

```
In[86]:= Clear[W];W:=RandomReal[];W
Out[86]= 0.368729
```

Again we check.

```
In[87]:= W==W
Out[87]= False
```

The result is false since each time W is called, the function RandomReal is evaluated again. So the first W evaluates RandomReal, and the second W again evaluates RandomReal, even though they are the same symbol. The same approach is applicable to Rule ($\rightarrow$) and RuleDelayed (:>).

Code Performance

In Mathematica there are many ways to write an expression in the same form. However, when we carry out long code operations, it is possible that there is a better notation to improve the performance of the code and thus not consume too many computational resources. This can be achieved by the relative performance of different functions for the development of the same result. The Wolfram Language provides a measure of this. Timing function shows the performance in units of seconds to each process in relation to the value of \$TimeUnit, which is the CPU time it takes for the Wolfram Language kernel to carry out the process. \$TimeUnit is different and varies from system to system, so you might get something different—in my case, 1/1000.

In the following example, we will see how long it takes to calculate the expression $10^{100000000}$, which is 10 to the power of 100 million. Timing returns two values: the unit time and the result of the calculation, but we will suppress the output of $10^{100000000}$ because it is a very big value.

```
In[88]:= Timing[Power[10,100000000];]
Out[88]= {1.8125,Null}
```

```
In[89]:= Timing[10^100000000;]
Out[89]= {1.84375,Null}
```

As you see, there is a difference between each; this has to do with how the Wolfram Language processes each computation. To look at the absolute time of a computation, use AbsoluteTiming.

```
In[90]:= AbsoluteTiming[10^100000000;]
Out[90]= {2.09879,Null}
```

```
In[91]:= AbsoluteTiming[Power[10,100000000];]
Out[91]= {2.06765,Null}
```

Clearly there is a difference too, as in the case with Timing. To restrain a computation by time, use TimeConstrained; with this command, time constraints can be added to a calculation. If the code is still running and the time limit has been reached, the evaluation will be aborted. For example, abort the evaluation after 1 second has passed.

```
In[92]:= TimeConstrained[10^100000000,1]
Out[92]= $Aborted
```

Strings

Text can be useful when a description of the code is needed. Mathematica allows us to input text into cells and create a text cell that is related to the computations we are doing. Mathematica has different forms to work with text cells. Text cells can have lines of text, and depending on the purpose of the text, we can work with different formats of text like creating chapters, sections, or just general text. In contrast, to text cells, we can introduce comments to expressions that need an explanation of their purpose or just a description. For that we simply write the comment within the symbols (* *). And the comments will be shown with different color; comments also always remain as unevaluated expressions. Comments can be single-line or multi-line.

Mathematica has the capability to work with strings. To introduce a string, just type the text enclosed in quotation marks "text", then Mathematica will know that is dealing with text. The characters can be whatever we type or enter into the cells.

```
In[93]:= "Hello World" (* This is a comment*)
Out[93]= Hello World
```

Mathematica assumes that what we enter is text by being enclosed in quotation marks, although we can always impel it to explicitly treat it as text using the ToString command. And we can check the head of the expression to make sure we are dealing with strings.

```
In[94]:= ToString[23.423563]
Out[94]= 23.4236
```

```
In[95]:= %//Head(* We use Head to knwo what type of objetc is *)
Out[95]= String
```

Strings appear without apostrophes when entered; this is because it is the default format.

```
In[96]:= "Welcome to Mathematica"
Out[96]= Welcome to Mathematica
```

Whenever we put the type cursor over a string in Mathematica and enter input, it will automatically appear surrounded by apostrophes. In this way, we can know we are working with strings.

As seen in the functionality of AtomQ, we can demonstrate that strings cannot have subexpressions in the Wolfram Language.

```
In[97]:= AtomQ["The sky is blue and tomorrow is expected to rain"]
Out[97]= True
```

Separates a string by characters.

```
In[98]:= Characters["Hello World"] (*Function that breaks the string into
its characters*)
Out[98]= {H,e,l,l,o, ,W,o,r,l,d}
```

Replace particular characters in a string with a rule operator.

```
In[99]:= StringReplace["Hello this is a string ", {"h","H"}→"4"]
(*This function repalce the string each time it appears for rule of the
pattern, that is 4*)
Out[99]= 4ello t4is is a string
```

Convert a text string to uppercase.

```
In[100]:= ToUpperCase["hello my name is"]
Out[100]= HELLO MY NAME IS
```

Now to lowercase.

```
In[101]:= ToLowerCase["HELLO MY NAME IS"]
Out[101]= hello my name is
```

Join a text string.

```
In[102]:= StringJoin["Nice","to","have","you","back"]
Out[102]= Nicetohaveyouback
```

Or with the string join symbol <>, which is StringJoin.

```
In[103]:= "Nice"<>"to"<>"have"<>"you"<>"back"
Out[103]= Nicetohaveyouback
```

How Mathematica Works

How Computations are Made (Form of Input)

Each time Mathematica receives a computation in the input cell, it uses the StandardForm, which is the output representation of expressions in the Wolfram Language, and has many aspects of common mathematical notation. Input can be written in various forms, but to know how the expression is written in the Wolfram Language, StandardForm is used.

```
In[104]:= StandardForm[1/x + x^2]
```
$$\text{Out[104]//StandardForm} = \frac{1}{x} + x^2$$

Now InputForm works similar but produces the output acceptable to be entered as Wolfram Language input.

```
In[105]:= {InputForm[ 1/x + x^2 ],InputForm[a^x],InputForm[a_x],InputForm √2 }
Out[105]= {x^(-1) + x^2,a^x,Subscript[a, x],Sqrt[2]}
```

Clearly every type of format has its equivalent in one line of code text, like the square that the symbol ($\sqrt{}$)means the same as Sqrt[ ]. To convert input into StandardForm, InputForm, and other forms, select the cell block and head to Cell ➤ Convert To ➤ StandardFrom, and InputForm, among others. StandardForm and InputForm apply to every expression in the Wolfram Language. Try using InputForm on the plots we made before to see how the expression is written completely. To understand better how Mathematica works; we want to know how symbolic or numeric computations are performed or written, the commands FullForm and TreeForm can be applied to view how expressions are represented symbolically. TreeForm represents the command in a graphical format, while FullForm represents the form of the expression managed by the Wolfram Language internally.

```
In[106]:= FullForm[ t/2 +2^2 ]
Out[106]//FullForm= Plus[4,Times[Rational[1,2],t]]
```

FullForm also represents the input as a one-line output code like InputForm. But even if InputForm also returns a one-line output code, why not use InputForm? The reason is because FullForm represents what Mathematica understands as input. With this in mind, FullForm is useful because it allows you to know what Mathematica interprets about the input that is written. In Mathematica the mathematical order of operations is preserved. So the previous output is as follows: first Mathematica detects the the rational number 1/2 (Rational[1,2]) and the symbol t followed by the multiplication of these two elements (Times[Rational[1,2],t]) followed by the addition of 2^2 (Plus[4,Times[Rational[1,2],t]]).

Another type of command that helps in creating a visualization of how Mathematica manipulates expressions is TreeForm. TreeForm returns the expression as a tree plot (Figure 1-28). As an alternative, we can apply commands with the post fix form "expr" // function. Rather than writing in the canonical form F ["expression"].

```
In[107]:= t/2 +2^2 //TreeForm

Out[107]//TreeForm
```

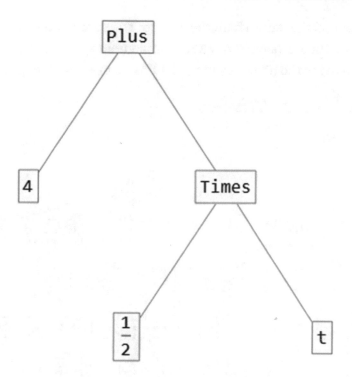

Figure 1-28. *Tree plot representation*

In short terms, Mathematica detects the multiplication of 1/2 times t and then proceeds to add the result of the product with the result of two squared. The tree graph is read from bottom to top, until you reach the top of the tree.

One more helpful command is Trace. Trace returns individual forms that correspond to the evaluation line, which contains the sequence of forms of the evaluated expression.

In[108]:= Trace[Plus[4,Times[Rational[1,2],t]]]

Out[108]= {{{Rational[1,2], $\frac{1}{2}$ }, $\frac{t}{2}$ }, 4+$\frac{t}{2}$ }

So here the sequence of operations are first the term Rational[1, 2], then 1/2, then 1/2 is multiplied by t and the result added to 4. Using FullForm in Trace lets you see how the internal structure changes.

In[109]:= FullForm[Trace[Plus[4,Times[Rational[1,2],t]]]]
Out[109]//FullForm=
List[List[List[HoldForm[Rational[1,2]],HoldForm[Rational[1,2]]],HoldForm
[Times[Rational[1,2],t]]],HoldForm[Plus[4,Times[Rational[1,2],t]]]]]

It can be seen that the terms change each step. The command HoldForm is used to see the output in an unevaluated form. As a complement to Trace, FullForm TreeForm can be combined to see the hierarchy of operations in an expression internally.

$$\text{In[110]:= Trace}\left[\frac{t}{2}+2\wedge2\right]\text{//TreeForm}$$

Out[110]//TreeForm

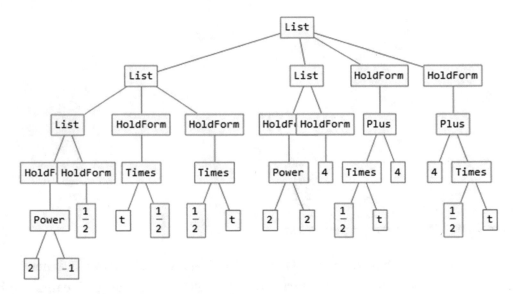

Figure 1-29. *TreeForm and Trace combined*

Here the tree (Figure 1-29) shows how changes are made and read from left to right. Reading the tree, we see that Mathematica recognizes that 1/2 is 2^-1; this is followed by t times 1/2, followed by 2^2, which is 4, and so on until the end. Moving the cursor over block will display the representation of the operation being held. It is necessary to emphasize that there may be occasions where we come across operations or expressions that we do not understand. A solution to this would be to use the previous commands as they allow you to see the inner structure of the expression and thus be able to understand how the operation is performed.

Searching for Assistance

The Wolfram Documentation Center contains the registry of all built-in functions. Documentation of functions can be accessed through the front end by opening a new window, clicking the Help tab on the toolbar, or by input expressions. The Input Assistant is displayed as an autocomplete or suggestion bar, when a command or related sensible options are written. When writing a built-in function or command, Mathematica will try to automatically complete the phrase.

Like in Figure 1-30, we type the word Random, and different commands associated with Random appear as suggestions. If the desired command is listed, we can select it with the cursor pointer.

Figure 1-30. *Autocomplete pop-up menu*

To access the documentation for a particular command, click the i document icon next to the command name, and the documentations windows should appear.

Note   Autocomplete also works for assigned symbols.

As you notice when writing the built-in function or command followed by the left square bracket, the completion menu appears; if you click on double-down arrow, it will display the inputs forms supported by that command, as shown in the Figure 1-31.

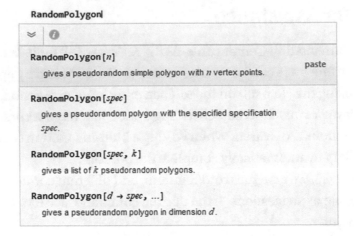

Figure 1-31. *Built-in function RandomPolygon with different input forms*

As seen in the example, the function RandomPolygon has four types of input form; also in the menu, we can see text related to the different forms of the input.

Getting how to know a function works, or how built-in functions are written. The best resource is to consult the Wolfram Documentation Center, as an alternative input expression can be used. So if we need help understanding how the function Head works, we simply input a question mark (?) before the name of the function, and it gives us a simple understanding of how the command works (Figure 1-32). If we want additional information related to the attributes of the function, a double question mark (??) can be employed. As a piece of advice, the Wolfram Documentation Center can be used for more in-depth options.

```
In[111]:= ?Head
Out[111]=
```

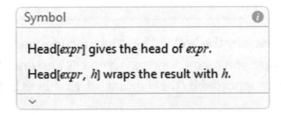

Figure 1-32. *Output information for the Head Command*

In the previous command, we showed how to show information related to a specific function, but if we don't recall the exact spelling, we can write the first letters of the name followed by an asterisk (*), and Mathematica gives us the list that matches our query. In the following example, the output are the functions that name start with "Hea". In the scenario that we needed more in-depth knowledge, the Wolfram documentation can be used.

```
In[112]:= ?Hea*
Out[112]=
```

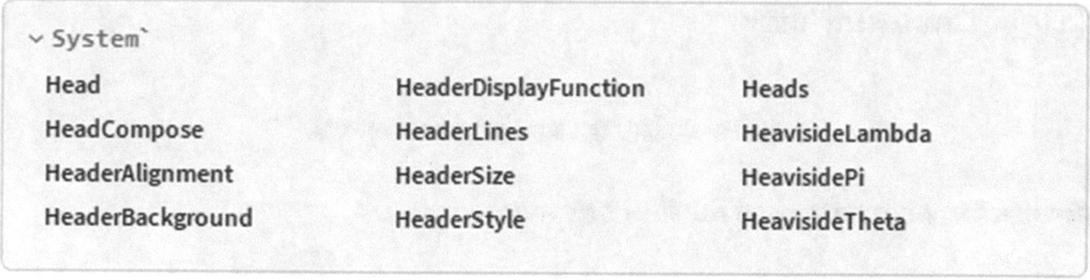

∨ System`		
Head	HeaderDisplayFunction	Heads
HeadCompose	HeaderLines	HeavisideLambda
HeaderAlignment	HeaderSize	HeavisidePi
HeaderBackground	HeaderStyle	HeavisideTheta

Figure 1-33. *Output information for the commands starting with the letters Head*

Handling Errors

Mistakes may be commonplace, as you most commonly develop code as you continue to learn. When a function fails, Mathematica will display a message below the written function. The message form provides the name of the function associated with the error along with a possible description of the cause of the error.

Next, we will see how this is.

```
In[113]:= StringJoin["hello","I am ",Jeff]
Out[113]= helloI am <>Jeff
```

⋯ StringJoin: String expected at position 2 in helloI am <>Jeff.

Figure 1-34. *Error message for the code entered*

The associated function in the message appears in red (Figure 1-34).What happens here is that the StringJoin function works only for strings and we are writing a Jeff variable, not a string, hence the error.

To find more about the error, we must click on the red ellipsis icon. A menu will appear listing the different options available to handle the error. To review the error in the documentation, we must click on the error option, which is the option that has an open book icon. This option will take us to the documentation of the associated function.

Another option from the pop-up menu that appears is Show Stack Trace. This is an option that shows us graphically and in blocks how the function and its expressions are being evaluated. This option is analogous to the Trace command we saw earlier.

Let us see the next example.

```
In[114]:= Power[x/0,2]
Out[114]= ComplexInfinity
```

$$\cdots \; \textbf{Power: } \text{Infinite expression } \frac{1}{0} \text{ encountered.}$$

Figure 1-35. *Error message for the code entered*

Here the error (Figure 1-35) is that Mathematica encounters a division by zero, which is undefined, and we can see the trace of the function with Stack Trace in Figure 1-36.

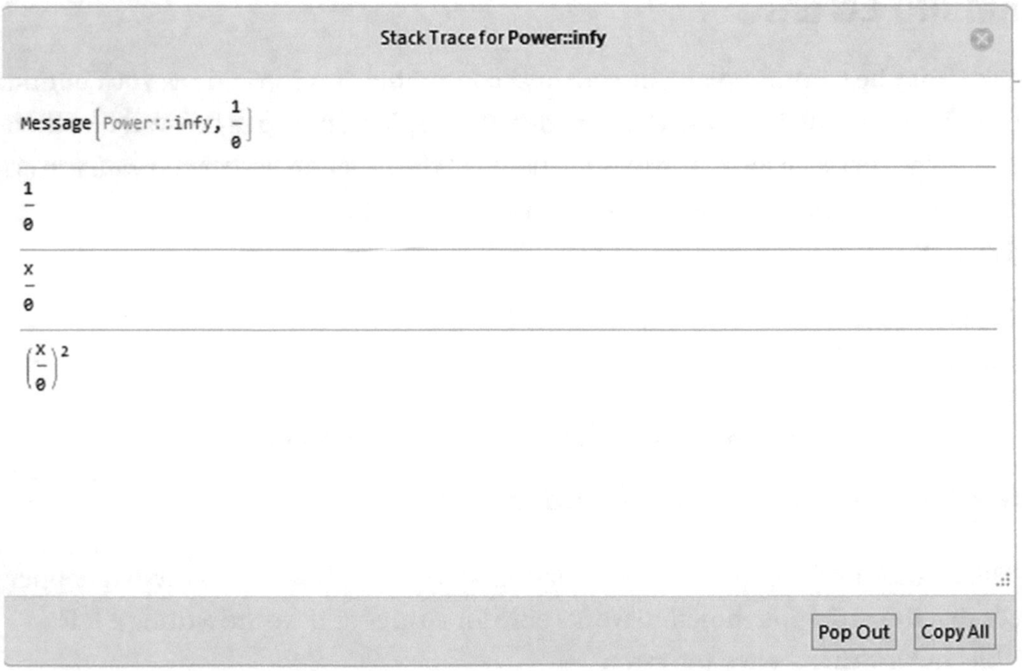

Figure 1-36. *Show Trace Stack pop-up window*

Notebook Security

The Wolfram Language provides creation and the ability to run dynamic content. These contents allow the user to create programs that can perform useful and complex tasks; on certain occasions, there may be the possibility of unwanted content being executed or code misuse. Now a notebook may or may not contain dynamic content as part of its code. Notebooks containing dynamic content can be instantly downloaded without any user action. Sometimes Mathematica alerts the user when a notebook contains dynamic content, displaying a message like that shown in Figure 1-37.

Figure 1-37. *Warning message of dynamic content*

If the notebook is not found in a trusted directory, the message will appear warning the user that the notebook contains unreliable dynamic content, as in Figure 1-37 .If the notebook is located in a reliable directory, the dynamic content will be executed without displaying a previous message to the user. To find out if a notebook is located in a trusted directory with the name TrustedPath, check out the trusted math directories, which are found in (1) $ BaseDirectory, (2) $ UserBaseDirectory, and (3) $ InitialDirectory.

```
In[115]:= $BaseDirectory
Out[115]= C:\ProgramData\WolframDesktop

In[116]:= $UserBaseDirectory
Out[116]= C:\Users\My-pc\AppData\Roaming\WolframDesktop

In[117]:= $InitialDirectory
Out[117]= C:\Users\My-pc\Documents
```

These are the trusted directories in my case; yours can defer from mine. By default the directories called UntrustedPath are those from which you can store files that can be potentially harmful, such as files downloaded from the internet. For this, in the Wolfram Language, the user's writing directories and the user's configuration directory are called UntrustedPath. To add, change, or remove the trusted and untrusted directories, go to the menu Edit ➤ Preferences, and then to the Security tab. There will be the options to edit the unreliable and trusted directories.

CHAPTER 2

Data Manipulation

In this chapter, we will review the basics of data creation and data handling in the Wolfram Language. The chapter begins with the concept of lists; we define what can be included within this structure as well as the creation of lists, nested lists, arrays, vectors, and matrices. We explore how to order a list, how to assign new values, and finally how to select elements of a list depending on an established pattern.

Lists

Lists are the core of data construction in the Wolfram Language. Lists can be used to gather objects, construct data structures, create tables, store values or variables, make elementary to complex computations, and for data characterization. As an overall, a list can represent any expression in the Wolfram Language (numbers, text, data, images, graphics, etc.)—that is, any set of whichever data.

If we access the information of List, we can see in Figure 2-1 that the common structure of how to form a list. Lists are represented by either curly braces or with the List command. In the Wolfram Language, almost every result data object can be listable; in other words, lists allow us to group data that maintain some type of relationship, even if they are of a different type, by manipulating all together (using the same identifier) or each one separately.

```
In[1]:= ??List
Out[1]=
```

© Jalil Villalobos Alva 2021
J. Villalobos Alva, *Beginning Mathematica and Wolfram for Data Science*,
https://doi.org/10.1007/978-1-4842-6594-9_2

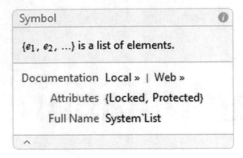

Figure 2-1. *List definition in the Wolfram Language*

As can be seen in the evaluation, elements are separated by commas, and the whole list is in between curly braces. Also, List is a protected variable; this means that we cannot assign values to the name List.

Types of Numbers

The fundamental number types in the Wolfram Language are those that are represented by integers, rational, real, and complex numbers.

First the integers have an exact result since they are numbers that cannot be represented by a decimal point.

```
In[2]:= {10, InputForm[10]}
Out[2]= {10,10}
```

Therefore, integers in the Wolfram Language are handled with infinite precision and infinite accuracy.

```
In[3]:= {10//Accuracy, InputForm[10]//Precision}
Out[3]= {∞, ∞}
```

Second, rational numbers are those that can be represented as a quotient of two integers.

```
In[4]:= {5/10,InputForm[10/12]}
Out[4]= { 1/2 ,5/6}
```

Mathematica treats rational numbers exactly as with integers, that is why whenever Mathematica deals with rational numbers, it returns as a result the minimum expression in which that number is represented.

```
In[5]:= {5/10 //Accuracy,InputForm[10/12] //Precision}
Out[5]= {∞  ∞}
```

Third, there are the real numbers, commonly known as float point numbers. In the Wolfram Language, any number that contains a decimal point is represented as a real number.

```
In[6]:= {2.72 //Precision, InputForm[2.72]}
Out[6]= {MachinePrecision,2.72}
```

Since the real numbers are approximate, therefore they do not have an exact precision, but they are numbers considered as machine numbers, which have the precision of the $MachinePrecision variable. It should be noted that in the Wolfram Language, numbers 1 and 1.0 are treated differently. Although Mathematica recognizes that they are equivalent expressions, it must be taken into account, that they are not the same object within the Wolfram Language.

To corroborate this, let's look at the following example, where we use SameQ to test if the expressions are the same for 1 and 1.0.

```
In[7]:= SameQ[Head[1],Head[1.0]]
Out[7]= False
```

We see that the heads of the expressions are different; that is because one is an integer and the other a real number.

```
In[8]:= {Head[1],Head[1.0]}
Out[8]= {Integer, Real}
```

Complex numbers are those numbers that contain a real part and an imaginary part. The form of a complex number is a + bi, in which the a term belongs to the real part and the term b to the imaginary part. The symbol "i" in mathematics represents the square root of the negative number - 1.

```
In[9]:= 10+19I
Out[9]= 10+19I
```

The type of precision in these numbers can be exact or approximate since these numbers can be built from the numbers described previously.

```
In[10]:= {precision[I], precision[1/2+0.3I], FullForm[11+1I]}
Out[10]= {∞ Machineprecision, Complex[11, 1]}
```

We know that complex numbers are atomic expressions, but even so, these numbers can be subdivided into different expressions, such as the case when we extract the real part and the imaginary part of each number.

```
In[11]:= 1+I //AtomQ
Out[11]= True
```

```
In[12]:= {ReIm[1+3I],Re[1+0.3I],Im[Complex[1,0.2]]}
Out[12]= {{1,3},1.,0.2}
```

When we deal with transcendental numbers like pi and the golden ratio, these numbers are treated as symbols—that is, Mathematica has reserved these symbols since they are important numerical constants, and therefore they have an exact precision despite being real numbers.

```
In[13]:= {Accuracy[\[Pi]],Precision[E],Accuracy[I],Precision[GoldenRatio]}
//NumberQ
Out[13]= False
```

To know whether or not a number is considered a number within the Wolfram Language, the NumberQ command gives us a result True if the expression is a number and False if it is not.

```
In[14]:= {NumberQ[1/2],NumberQ[1],NumberQ[E]}
Out[14]= {True,True,False}
```

As a result, we can see how a rational number and an integer are numbers, but the number e is not. In fact, we can see that you are of type symbol.

```
In[15]:= {Head[E],FullForm[E]}
Out[15]= {Symbol,E}
```

Generally speaking, there is no restriction on combining the different types of numbers within the Wolfram Language. We can perform operations between different types.

```
In[16]:= {1+0.2+1/2+1+11+1I}
Out[16]= {13.7 +1. I}
```

Conversion between approximate numbers to exact numbers is carried out with Rationalize.

```
In[17]:= Rationalize[2.72]
Out[17]= 68/25
```

Also, a different number notation is supported, such as scientific notation; scientific notation is a very useful tool to represent large numbers in powers of ten.

```
In[18]:= ScientificForm [ N@e/100000 ]
Out[18]//ScientificForm= 2.71828x10-5
```

We know that the function N is used to calculate approximate numbers. It converts the exact expression to an approximate one, remembering that the desired precision can also be included.

There are also different forms that can generally be extrapolated to all the built-in functions notations of the Wolfram Language.

1) Employing the direct application of the function N [ ] to the expression.

   ```
   In[19]:= N[13/7]
   Out[19]= 1.85714
   ```

2) Utilizing the infix notation, ~N~.

   ```
   In[20]:= e~N~e
   Out[20]= 2.72
   ```

3) Through the postfix notation, // N.

   ```
   In[21]:= e//N
   Out[21]= 2.71828
   ```

4)   By means of the prefix notation, N@.

```
In[22]:= N@e
Out[22]= 2.71828
```

When the precision is not defined, Mathematica uses the value of $MachinePrecision to determine the standard precision of the approximate number. The value of $MachinePrecision varies since it is a float number established by Mathematica according to the characteristics of each computer.

```
In[23]:= $MachinePrecision
Out[23]= 15.9546
```

Setting arbitrary precision with SetPrecision:

```
In[24]:= SetPrecision[e,17]
Out[24]= 2.7182818284590452
```

Using machine precision:

```
In[25]:= SetPrecision[e,MachinePrecision]
Out[25]= 2.71828
```

When precision is not introduced Mathematica will use MachinePrecision numbers.

```
In[26]:= SetPrecision[e,MachinePrecision] =N@e
Out[26]= True
```

Another way to enter approximate numbers with some precision is by entering the grave accent symbol (`) after the real number followed by the precision. Here we use it for six-digit precision.

```
In[27]:= 77/3`6
Out[27]= 25.6667
```

Working with Digits

To extract digits that make up an exact number, use the IntegerDigits function.

```
In[27]:= IntegerDigits[234544553]
Out[27]= {2,3,4,5,4,4,5,5,3}
```

RealDigits for approximate numbers:

```
In[28]:= {RealDigits[321.4546554], RealDigits[N@e]
Out[28]= {{{3,2,1,4,5,4,6,5,5,4,0,0,0,0,0,0},3},{{2,7,1,8,2,8,1,8,2,8,4,5,
9,0,4,5},1}}
```

In the case of a complex number, it would consist of extracting its real and imaginary parts and then proceeding to extract the digits of each part, as the case may be.

```
In[29]:= RealDigits[ReIm[113+2.7213I]]
Out[29]= {{{1,1,3,0,0,0,0,0,0,0,0,0,0,0,0,0,0},3},{{2,7,2,1,3,0,0,0,0,0,0,0,0,
0,0,0,0},1}}
```

By default, the two previous functions give results in the decimal base. To define a base, just enter the base you want as the second argument of the function.

```
In[30]:= RealDigits[321.4546,2]
Out[30]= {{1,0,1,0,0,0,0,0,1,0,1,1,1,0,1,0,0,0,1,1,0,0,0,0,0,1,0,1,0,1,0,1,
0,0,1,1,0,0,1,0,0,1,1,0,0,0,0,1,1,0,0,0,0},9}
```

In the latter example, we extract the digits of the number on base 2.
Specification on showing 3 digits on the number e in base 10.

```
In[31]:= RealDigits[N@e, 10, 3]
Out[31]= {{2,7,1},1}
```

To reconstruct a number from the representation of their integers is possible with the function FromDigits.

```
In[32]:= FromDigits[{2,7,1,1}]
Out[32]= 2711
```

Also, it is possible to form a float point number,

```
In[33]:= N@FromDigits[{{2,7,1,1},1}]
Out[33]= 2.711
```

and to measure the length of an integer number.

```
In[34]:= IntegerLength[2711]
Out[34]= 4
```

A Few Mathematical Functions

The Wolfram Language has a wide repertoire of mathematical functions from the most basic to the most specialized. These functions can be handled numerically or symbolically, which allows for pure analytical manipulation.

Trigonometric functions are available either in radians or in degrees. Typing a number alone will calculate and return the value in radians.

```
In[35]:= Cos[Pi]
Out[35]= -1
```

Entering the number followed by the unit Degree or the symbol of degrees (°) will calculate and return the value in degrees.

```
In[36]:= Sin[90 Degree]==Sin[90\[Degree]]
Out[36]= True

In[37]:= Sin[90\[Degree]]
Out[37]= 1
```

The same is applicable for hyperbolic trigonometric functions and inverse trigonometric functions.

```
In[38]:= N[Cosh[Pi]]
N[Tanh[45 Degree]]
Out[38]= 11.592
Out[39]= 0.655794

In[40]:= N[ArcTan[Pi]]
N[ArcSinh[45 Degree]]
Out[40]= 1.26263
Out[41]= 0.721225
```

Logarithmic functions and exponential functions are written like common math notation. Logarithms with only a number will compute the natural logarithm.

```
In[42]:= Log[E]
Out[42]= 1
```

To specify a base, type the number as first argument, and the base as second argument.

```
In[43]:= Log[10,10]
Out[43]= 1
```

Exponentials can be written with Exp or with the constant e.

```
In[44]:= Exp[2]==E^2
Out[44]= True
```

The factorial is represented by either typing the exclamation mark after the number or by using Factorial.

```
In[45]:= 12!
Out[45]= 479001600
```

```
In[46]:= Factorial[12]
Out[46]= 479001600
```

Numeric Function

In the Wolfram Language, there are functions that provide manipulation of numerical data. These functions can work with any of the types of numbers (real, integer, rational, complex). The precision can be handled completely—that is, a number in its most exact form or with floating point precision.

To truncate a number z closest to z, use Round with no arguments. Adding a second argument rounds the number z to the nearest multiple of the second number.

```
In[47]:=Round[8.9](*Rounds to 9 because it is the closest number*)
Out[47]= 9
```

```
In[48]:=Round[8.9,2](*Rounds to 8 because it is the closest multiple of 2, 2^3*)
Out[48]= 8
```

Other similar functions that can truncate numbers given a number z are the Floor and Ceiling. The first one rounds to the largest integer less than or equal to number typed, and the second one rounds to the smallest integer larger than or equal to the typed number.

```
In[49]:= Floor[Pi]
Out[49]= 3
```

```
In[50]:= Ceiling[Pi]
Out[50]= 4
```

Floor and Ceiling functions can be written in their mathematical notation, $\lfloor z \rfloor$ for Floor and $\lceil z \rceil$ for Ceiling, by typing the key ESC lf ESC for left Floor and ESC rf ESC for right Floor. The same is true for Ceiling—just change lf for lc (left Ceiling) and rc (right Ceiling).

```
In[51]:= ⌊Pi⌋
Out[51]= 3
```

```
In[52]:= ⌈Pi⌉
Out[52]= 4
```

Converting a float point number to a rational approximation can be done with Rationalize. But adding the number 0 as the second argument can force the calculation to find the most exact form of a float point number. For example, a rational approximation to the number e.

```
In[53]:= Rationalize[N[E],0]
Out[53]= 325368125/119696244
```

The functions Max and Min return the maximum and minimum number of a list of numbers.

```
In[54]:= Max[9,8,7,0,3,12]
Out[54]= 12
```

```
In[55]:= Min[0987,32,9871]
Out[55]= 32
```

Lists of Objects

In this section, we will extend the concept of list in the Wolfram Language to techniques for creating lists, creating nested lists through specialized functions, and how to effectively store data in a variable. The theme is developed from how to create data sets and how they can come from a wide range of functions, since the creation of a list can have a wide range of contents and those contents can be sets of numbers, text strings, equations, arithmetic operations, or any expression in Mathematica. Despite this, we will see concepts such as arrays, sparse arrays, and their respective object types. We will also cover the topic of nested lists and how you can create data in a nested form in various ways.

List Representation

The curly braces represent a list of general objects; the members of a list are separated by a comma. The simplest form to create a list is to enclose data in curly braces, or by using the function List. The following examples show how to assign lists to variables and how to gather objects into a list.

```
In[56]:= {x²+1, "Dog", π}
List[1,P,Power[3,2]] (* Power[3,2] represents 3 raised to the power of 2 *)
Out[56]= {x²+1, "Dog", π}
Out[57]= {1,P,9}
```

The list identifier or symbol is an optional name to create the structure.

```
In[58]:= List["23.22","Dog", π,2,4,6,456.,56,2==3 && 3==2]
Out[58] = {23.22,Dog, π,2,4,6,456.,56,False}
```

Inside a list, between the braces, we can define all the elements that we consider suitable to be listed.

```
In[59]:= {1+I, π + π,"number 4",Sin[23 Degree],425+I-413-3I,24,4456.,
"dog"+"cat"}
Out[59]= {1+I, 2π,number 4,Sin[23 °] 12-2 I,24,4456.,cat+dog}
```

In Mathematica, there are different types of objects. To identify an object type, we have to use the Head function. The returning value is the head of the expression, known as the data type. If we apply Head to a list, we would get that the head of the expression is a list.

```
In[60]:= % //Head
Out[60]= List
```

This means that the object we have created is a List object.

Generating Lists

Lists can be created with costume values, but Mathematica has a variety of functions to create automated lists, like Range and Table. Range and Table functions create an equally spaced list of numbers. But Table generates a list with specified intervals; the interval of the table specifies that "i" goes from 1 to 10. Wolfram Language also allows us to include built-in functions inside a list.

```
In[60]:= Range[10]
Table[i,{i,1,10}]
Table["Soccer",{i,1,15}]

Out[60]= {1,2,3,4,5,6,7,8,9,10}
Out[61]= {1,2,3,4,5,6,7,8,9,10}
Out[62]= {Soccer,Soccer,Soccer,Soccer,Soccer,Soccer,Soccer,Soccer,Soccer,
Soccer,Soccer,Soccer,Soccer,Soccer,Soccer}
```

Table also can be used to create indexed lists. Each interval is specified within the curly braces { }. Multiple examples are shown.

```
In[63]:= Table["Red and Blue", 5]
Range[-5,5]

Out[63]= {Red and Blue,Red and Blue,Red and Blue,Red and Blue,Red and Blue}
Out[64]= {-5,-4,-3,-2,-1,0,1,2,3,4,5}
```

Table can work with or without an inner iterator, but in order to create structured lists, we recommend using an iterator.

```
In[65]:= Table[i^i,{i,1,5}]
Out[65]= {1,4,27,256,3125}
```

This shows the function without an iterator.

```
In[66]:= Table[10^3,{5}]
Out[66]= {1000,1000,1000,1000,1000}
```

Note   When using the iterator, make sure to properly write the expression to avoid error. When Table recognizes the iterator, it changes colors, because the letter used is no longer a symbol.

We can create a list of lists. Later we will see that this type of structure is considered a nested list.

```
In[67]:= {Range[5],Table[h,{h,-6,2}]}
Out[67]= {{1,2,3,4,5},{-6,-5,-4,-3,-2,-1,0,1,2}}
```

The iterator can be also an alphanumeric variable.

```
In[68]:= Table[data2,{data2,0,6}]
Out[68]= {0,1,2,3,4,5,6}
```

Structures of arrays of the same data can also be created, such as an array of 2 x 2.

```
In[69]:= Table[11,{2},{2}]
Out[69]= {{11,11},{11,11}}
```

Table supports more multiple iterators. This is useful when trying to construct tabular data.

```
In[70]:= Table[i+j+k,{i,1,4},{j,1,4},{k,1,4}]
Out[70]= {{{3,4,5,6},{4,5,6,7},{5,6,7,8},{6,7,8,9}},{{4,5,6,7},{5,6,7,8},
{6,7,8,9},{7,8,9,10}},{{5,6,7,8},{6,7,8,9},{7,8,9,10},{8,9,10,11}},
{{6,7,8,9},{7,8,9,10},{8,9,10,11},{9,10,11,12}}}
```

To display a list in a more structured way using the command Grid:

```
In[71]:= Table[i-j,{i,1,2},{j,1,2}]//Grid
Out[71]=  0  -1
          1   0
```

An alternative to Grid is TableForm. With TableForm we can display the list created as a table. Later, we will see how to use the command TableForm more in-depth.

```
In[72]:= Table[i+j,{i,1,2},{j,4,6}] //TableForm
Out[72]//TableForm=
5  6  7
6  7  8
```

There is no limitation on the intervals of the iterators. We can choose that "i" goes from 0 to 3 and j from "i" to 3 and use TableForm to view it.

```
In[73]:= Table[{i, j}, {i, 3}, {j, i, 3}] // TableForm
Out[73]//TableForm=
 1   1   1
 1   2   3
 2   2
 2   3
 3
 3
```

We can even use other syntax notations like the increment (++) or decrement (--) in the interval iterator.

```
In[74]:= Table[{i, j},{i,2},{j,i++,2}]
Out[74]= {{{2,1},{2,2}},{{3,2}}}
```

The increment (++) and decrement (--) operators can also be used in assigned variables; this operator can also have precedence or posteriority. When written before the variable, they are called PreIncrement or PreDecrement, respectively.

```
In[75]:= x=0;x++;x (*applied on the current value and shown next time x is
called*)
Out[75]= 1
In[76]:= Clear[x];x=0;--x (*applied on the current value and shown when x
is called*)
Out[76]= -1
```

As another alternative, we perform replacement rules with the symbol (/.). For example, we generate a list of random integers from 0 or 1 and then replace the 1's with

2's whenever they appear. Make sure to add a space between the condition expressions to avoid a typo error.

```
In[77]:= Table[RandomInteger[],{i,1,10}]/. 1 → 2
Out[77]= {2,0,2,2,2,0,0,0,2,2}
```

Arrays of Data

There are different forms to create an array. The most used form is a list like we saw in the previous section. But as an alternative to the command Table or Range, arrays can be created with the command Array. What Array generates is a list with a specific function applied to the elements created.

In addition to the functions already mentioned, other functions can be used to build lists, like Array, ConstantArray, and SparseArray. The form of these functions is analogous to the previous ones.

```
In[78]:= Array[Cos[90 Degree],{3,3}]//Grid
Out[78]=     0[1,1] 0[1,2] 0[1,3]
             0[2,1] 0[2,2] 0[2,3]
             0[3,1] 0[3,2] 0[3,3]
```

What happens with Array is that it constructs an array from a function. In the previous example, we generated an array from the numerical value of the cosine of 90 degrees, followed by the structure of the array, which is 3 x 3. The indices that appear on the right side of the array values are the positions of each element in the array.

If we generalize to any function, we can better see how Array works.

```
In[79]:= Array[F,{2,2}] //Grid
Out[79]=     F[1,1] F[1,2]
             F[2,1] F[2,2]
```

As we can observe, the function F is applied and is respective to each element of the arrangement. To create an array of constant values the ConstantArray function is used. To write the function, we first write the value we want to repeat followed by the times we want it to repeat.

```
In[80] := ConstantArray[π,5]
Out[80] = {π, π, π, π, π]}
```

We can also create arrangements with defined dimensions.

```
In[81]: = ConstantArray[π,{4,4}]
Out[81] = { { π, π, π, π },{ π, π, π, π },{ π, π, π, π },{ π, π, π, π } }
```

To display a data array, there is the MatrixForm command, which, as its name suggests, shows the array in matrix form.

```
In[82]: = ConstantArray[π,{4,4}] //MatrixForm
Out[82]//MatrixForm=
```

$$
\begin{pmatrix}
\pi & \pi & \pi & \pi \\
\pi & \pi & \pi & \pi \\
\pi & \pi & \pi & \pi \\
\pi & \pi & \pi & \pi
\end{pmatrix}
$$

A sparse arrangement is an arrangement in which the elements generally have the same value. With SparseArray we can define the values of the array positions. By standard, if any position is not defined, the value will be 0.

```
In[83]:= SparseArray[{{1,1},{2,2}} → {1,2}]
Out[83]=
```

Figure 2-2. *SparseArray Object*

The result of a SparseArray generates an object of type SparseArray, which is shown in Figure 2-2, with the name of the command and a gray box that appears. If you click on the + icon, you will see the characteristics of the array as well as its rules; this is shown in Figure 2-3.

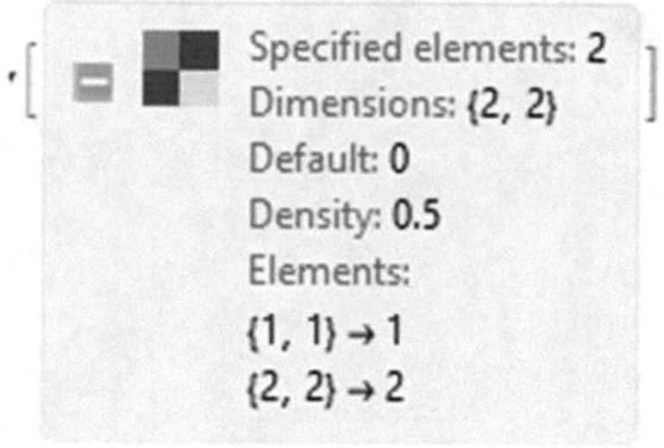

Figure 2-3. *Specifications of the array*

In the Wolfram Language, there is no limitation on the content of a SparseArray. Furthermore, we can create an array in which we have the same values on its diagonal.

```
In[84]: = SpArray = SparseArray[{{1,1} → "A",{2,2} → "A",{3,3} →
"A",{4,4}→ "A" },{4,4}]
Out[84]=
```

Figure 2-4. *Sparse Array with more elements*

As you might notice in Figure 2-4, elements of the same values in the array appear in one color and different values appear in other color.

With the help of Matrix Form, we can visualize the arrangement as a Matrix.

```
In[85]:= MatrixForm[%]
Out[85]//MatrixForm=
```

$$\begin{pmatrix} A & 0 & 0 & 0 \\ 0 & A & 0 & 0 \\ 0 & 0 & A & 0 \\ 0 & 0 & 0 & A \end{pmatrix}$$

To convert the sparse array object to a list object, use Normal to normalize into expression form.

```
In[86]:= Normal[SpArray]
Out[86]= {{A,0,0,0},{0,A,0,0},{0,0,A,0},{0,0,0,A}}
```

And now we deal with a list.

```
In[87]:= Head[%]
Out[87]= List
```

Nested Lists

A nested list is a list of lists, where the elements of the lists correspond to another list, and so on. Nested lists can be used for ordered or unordered data structure. To create a nested list we can use curly braces within curly braces or built-in functions.

```
In[88]:= {{"This","is","A"},{"Nested","List","."}}
Out[88]= {{This,is,A},{Nested,List,.}}
```

We can also use the function Table.

```
In[89]:= Table[Prime[i]+Prime[j],{i,1,3},{j,2,4}]
Out[89]= {{5,7,9},{6,8,10},{8,10,12}}
```

To measure a list, we must use the Length command.

```
In[90]:= NestL=Table[Prime[i]+RandomReal[j],{i,1,3},{j,1,3}];
Length[NestL]
Out[90]= 3
```

As you might notice, the length of the list is 3, because Length is properly used with flattened lists. To properly measure the depth of a nested list, Dimensions is more suited for the task.

```
In[91]:= Dimensions[NestL]
Out[91]= {3,3}
```

With Dimensions we can get a general aspect of the dimensions of the nested list, the output generated, meaning that our list is constituted by a list of three sublists and that the sublists each have three elements. As we can see, Mathematica constructs a list that has three elements, in which those three elements are also a list, and those lists have three elements, and each element corresponds to a specific value.

Note   You might want to use TreeForm, where you can explore how Mathematica deals with nested list expressions. Try this code, for instance (*TreeForm[NestL]*).

A useful command to measure the depth of a nested list or an array is ArrayDepth.

```
In[92]:= ArrayDepth[NestL]
Out[92]= 2
```

Now we know programmatically that NestL has a depth of two.

Vectors

Mathematica will handle vectors the same way as with lists. Usual calculations of linear algebra can be symbolic or numeric.

```
In[93]:= V={6,3,2}
Out[93]= {6,3,2}
```

A vector is always shown as a list. To see a vector in regular notation, the command MatrixForm is used.

```
In[94]:= MatrixForm[V]
Out[94]//MatrixForm=
```

$$\begin{pmatrix} 6 \\ 3 \\ 2 \end{pmatrix}$$

The command VectorQ can tell us if the list we are dealing with is a vector or not.

```
In[95]:= VectorQ[V]
Out[95]= True
```

To see the rank of the vector, use either ArrayDepth or TensorRank.

```
In[96]:= {TensorRank[V],ArrayDepth[V]}
Out[96]= {1,1}
```

Vectors are created with the same commands that create a list, Table, Array, Range, curly braces, SparseArray, ConstantArray, etc. Also common operations of vectors are performed like normal lists.

```
In[97]:=
Print["Addition: "<>ToString[V+V]]
Print["Substraction: "<>ToString[V-V]]
Print["Scalar product: "<>ToString[2*V]]
Print["Cross product: "<> ToString[Cross[V,{1,3,2}]]]
Print["Norm: "<> ToString[Norm[V]]]
Ou[97]=
Addition: {12, 6, 4}
Substraction: {0, 0, 0}
Scalar product: {12, 6, 4}
Cross product: {0, -10, 15}
Norm: 7
```

Matrices

A matrix is a square list or list of lists, arranged in n-rows and m-columns, where n and m are the dimensions of the matrix.

$$\mathbf{A}_{m\times n}=\begin{pmatrix} a_{11} & a_{12} & \cdots & a_{1n} \\ a_{21} & a_{22} & \cdots & a_{2n} \\ \vdots & \vdots & \ddots & \vdots \\ a_{m1} & a_{m2} & \cdots & a_{mn} \end{pmatrix}$$

The easiest form is to create a list of lists.

```
In[98]:= {{3,3,1},{7,8,7}}//MatrixForm
Out[98]//MatrixForm=
```

$$\begin{pmatrix} 3 & 3 & 1 \\ 7 & 8 & 7 \end{pmatrix}$$

Another way is to go to Insert ➤ Table/Matrix ➤ New. A pop-up menu appears; within this menu select matrix and specify the rows and columns. With this option you can also specify to fill contents and the diagonal, as well as to add grid or frames, such as in the next example that has draw lines between columns.

$$\text{In[99]}:= \quad A = \begin{pmatrix} 1 & 0 & 0 \\ 0 & 1 & 0 \\ 0 & 0 & 1 \end{pmatrix}$$

```
Out[99]= {{1,0,0},{0,1,0},{0,0,1}}
```

To test whether a list of lists is a matrix, use MatrixQ.

```
In[100]:= MatrixQ[A]
Out[100]= True
```

Transpose returns the transpose of a matrix—that is, changing its rows by columns. For a matrix **A,** the transpose is denoted by $\mathbf{A}^T$.

```
In[101]:=Transpose[{{0,1,0},{0,1,0},{0,1,0}}]//MatrixForm
Out[101]//MatrixForm=
```

$$\begin{pmatrix} 0 & 0 & 0 \\ 1 & 1 & 1 \\ 0 & 0 & 0 \end{pmatrix}$$

Matrix Operations

Common operations between matrices are performed by the rules of linear algebra: addition, subtraction, and multiplication. Remember that when multiplying two matrices, **A** and **B,** the number of columns in **A** must match the number of rows in **B**. In mathematical terms: $\mathbf{A}_{m*n} \times \mathbf{B}_{n*l} = \mathbf{C}_{m*l}$.

```
In[102]:=
B={{0,1,0},{0,1,0},{0,1,0}};
Print["Addition: "<>ToString[A+B]]
Print["Substraction: "<>ToString[A-B]]
Print["Product: "<>ToString[Dot[B,V]]]
Out[102]=
Addition: {{1, 1, 0}, {0, 2, 0}, {0, 1, 1}}
Substraction: {{1, -1, 0}, {0, 0, 0}, {0, -1, 1}}
Product: {3, 3, 3}
```

To calculate the determinant, use Det.

```
In[103]:= {Det[A],Det[B]}
Out[103]= {1,0}
```

To construct a diagonal matrix, use DiagonalMatrix; for the identify matrix, use IdentityMatrix. DiagonalMatrix is for costume values, and the IdentityMatrix returns a matrix with a diagonal with the same elements.

```
In[104]:= DiagonalMatrix[{x,y,z}]//MatrixForm
IdentityMatrix[{2,2}]//MatrixForm(*Identity matrix of 2 by 2*)
Out[104]//MatrixForm=
```

$$\begin{pmatrix} x & 0 & 0 \\ 0 & y & 0 \\ 0 & 0 & z \end{pmatrix}$$

```
Out[105]//MatrixForm=
```

$$\begin{pmatrix} 1 & 0 \\ 0 & 1 \end{pmatrix}$$

Restructuring a Matrix

Matrix restructuring is done with the same commands to restructure a list, like replacing an element with a new value.

```
In[106]:= ReplacePart[A,{{1,1},{2,2}}-> 3]//MatrixForm
Out[106]//MatrixForm=
```

$$\begin{pmatrix} 3 & 0 & 0 \\ 0 & 3 & 0 \\ 0 & 0 & 1 \end{pmatrix}$$

Also, it can be done by assigning the new value. To access the elements of a matrix, enter the symbol followed by the subscript of the element of interest with the double bracket notation ([[ ]]). Later we will see the proper functionality of this short notation. In this case, we will change the value of the element in position 1,1 of the matrix.

```
In[107]:= A[[1,1]] = 2;
MatrixForm[A]
Out[107]//MatrixForm=
```

$$\begin{pmatrix} 2 & 0 & 0 \\ 0 & 1 & 0 \\ 0 & 0 & 1 \end{pmatrix}$$

If matrix A is called again, the new value will be preserved. To invert a square matrix, use Inverse.

```
In[108]:= Inverse[A]//MatrixForm
Out[108]//MatrixForm=
```

$$\begin{pmatrix} \dfrac{1}{2} & 0 & 0 \\ 0 & 1 & 0 \\ 0 & 0 & 1 \end{pmatrix}$$

Measuring the dimensions of a matrix can be done by using Dimensions.

```
In[109]:= Dimensions[A]
Out[109]= {3,3}
```

Manipulating Lists

In the previous section, we saw different ways of creating lists, either through arrays, nested lists, or tables. In this section we will go into detail on how to manipulate these lists through referenced names, functions, and compact notation. We will study how to access the data of a list depending on your position in it. We will see how to add and delete elements of a list, how to replace single parts, and how to change the value of a particular element. We will also learn how to restructure a list once it has been built, and we will see the ordering of a listand how to convert a nested list to a linear list, depending on how deep the list is. In addition, we will see how to see data from a list through patterns. We will study the pattern behavior in the Wolfram Language.

Retrieving Data

Several functions exist for handling elements of a list. The function Part ["list", i] allows you to select index parts of a list, with index i.

For example, let us define a list called list1 and use Part to access the elements inside the list. The Part function works by defining the position of the element we want

```
In[110]:= list1={1,2};
Part[{1,2},1]
Out[110]= 1
```

or with the index notation,

```
In[111]:= {1,2}[[1]]
Out[111]= 1
```

Lists can be fully referenced by using their assigned names. Elements inside the structure can be accessed using the notation of double square brackets [[i]] or with the special character notation of double brackets, "⟦ ⟧ ".

Tip   To introduce the double square bracket character, type Esc [[Esc and ESC]] ESC.

```
In[112]:= list1[[1]] (* [[ i ]] gives you access to the element of the list
in the postion i .*)
Out[112]= 1
```

Note   Square brackets ([[ ]]) are the short notation for Part.

To access the elements of the list by specifying the positions, we can use the span notation, which is with a double semicolon (;;).

```
In[113]:= list2=List[34,6,77,4,5,6];
Part[list2,1;;4] (* from items 1 to 4*)
Out[113]= {34,6,77,4}
```

We can also use backward indices, where the counts start from right to left, which is from the last element to the first. Let us now select from position -6 to -4.

```
In[114]:= list2[[-6;;-4]]
Out[114]= {34,6,77}
```

For the nested list, the same process is applied. The concept can be extended into a more general aspect. In the next example, we will create a nested list with three levels and select a unique element.

```
In[115]:= list3=List[2^3,2.72, {β, ex, {Total[1+2], "Plane"}}];
list3[[3,3,2]]
Out[115]= Plane
```

In the previous example, we create a nested list of depth three. Next we select the third element of the list, which is {8, 2.72, {β, ex, {Total[1 + 2], "Plane"}}, then from that list we select the three elements of the previous list, which is {Total[1 + 2], "Plane"}. Finally we select the element in second position of the last list, which is "Plane."

If we are dealing with a nested list, we use the same concept that we saw with the span notation. In the next example, we select the third element of the list3 and then display from position 1 to 2.

```
In[113]:=list3[[3,1;;2]]
Out[113]= {β, ex}
```

The same is done to a more in-depth list; we use the third element of the list, and then display from position 3 to 3 and select part 1.

```
In[114]:= list3[[3,3 ;; 3,1]]
Out[114]= {3}
```

Segments of data can be displayed based on what parts of the data we are interested in. For example, the function Rest shows the elements of the data except for the first. And Most reveals the whole list except for the last element(s), depending on the type of list.

```
In[115]:= Rest[list3]
Out[115]= {2.72, {β, ex, {3, Plane}}}
```

```
In[116]:= Most[list3]
Out[116]= {8,2.72}
```

An alternative to the previous functions is the function Take. With Take we can select more broadly the data in a list. There are three possible ways to accomplish this:

1. By specifying the first i elements.

   ```
   In[117]:= Take[list3,2]
   Out[117]= {8,2.72}
   ```

2. By specifying the last -i elements.

   ```
   In[118]:= Take[list3,-1]
   Out[118]= {{β, ex, {3, Plane}}}
   ```

3. By selecting the elements from i to j.

   ```
   In[119]:= Take[list3,{1,3}]
   Out[119]= {8, 2.72, {β, ex, {3, Plane}}}
   ```

Assigning or Removing Values

Once a list is established—that is, if we have defined a name for it—it can be used just like any other type. This means that elements can be replaced by others. To change a value or values, we select the position of the item and then we set the new value.

```
In[120]:= list4={"Soccer","Basketball",0,9};
list4[[2]]=1 (*position 2 corresponds to the string Basketball and we
change it for the number 1*)
Out[110]= 1
```

And we can check the new values have been added.

```
In[121]:= list4
Out[121]= {Soccer,1,0,9}
```

In addition to using the abbreviated abbreviation notation, we can use the function Replace part of specific values and choose the list, the new element, and the position.es.

```
In[122]:= ReplacePart[list4,Exp[x],4]
Out[122]= {Soccer,1,0, eˣ}
```

To add new values, we use PrependTo and AppendTo; the first adds the value on the left side of the list, whereas the second adds it by the right side of the list. Append and Prepend works the same but without storing the new value in the original variable.

```
In[123]:= PrependTo[list4,"Blue"]
Out[123]= {Blue,Soccer,1,0,9}
```

```
In[124]:= AppendTo[list4,4]
Out[125]= {Blue,Soccer,1,0,9,4}
```

```
In[126]:= list4(* we can check the addition of new values.*)
Out[126]= {Blue,Soccer,1,0,9,4}
```

To remove the values of the list, we use Drop. Drop can work with the level of the specification or the number of elements to be erased.

```
In[127]:= Drop[list4,3];(* first 3 elements, Delete[list3.3]*)
Drop[list4,{5}](* or by position, position, number 5*)
Out[127]= {Blue,Soccer,1,0,4}
```

The Delete command can also do the job by defining the particular positions on the list—for example, deleting the contents in positions 1 and 5.

```
In[128]:= Delete[list4,{{1},{5}}]
Out[128]= {Soccer,1,0,4}
```

As an alternative to Append and Prepend, there is the Insert function, with which we can add elements indicating the position where we want the new data. Inserting the expression (list4), the new element (2/43.23), the position (3rd position of the list). Now the number 2/43.23 is in the 3rd position in the list.

```
In[129]:= Insert[list4,2/43.23,3]
Out[129]= {Blue,Soccer,0.0462642,1,0,9,4}
```

Insert allows the use of several positions at the same time. For example, inserting the number 0.023 at positions -6 (2nd) and 7 (the las position).

```
In[130]:= Insert[list4,0.023,{{-6},{7}}]
Out[130]= {Blue,0.023,Soccer,1,0,9,4,0.023}
```

In the special case that we want to add repetitive terms or remove terms to a list or an array, we can use the ArrayPad function. If the term to be added is not defined, the standard value is zeros.

```
In[131]:= ArrayPad[list4,1](*number 1 means one zero each side*)
Out[131]= {0,Blue,Soccer,1,0,9,4,0}
```

In the case that we want to add one-sided terms, it is written as follows.

```
In[132]:= ArrayPad[list4,{1,2}](* 1 zero to the left and 2 zeros to the right*)
Out[132]= {0,Blue,Soccer,1,0,9,4,0,0}
```

To add values other than zero, we must write the value to the right of the number of times the value is repeated.

```
In[133]:= ArrayPad[list4,{0,3},"z"](*Adding the letter z three times only
the right side*)
Out[133]= {Blue,Soccer,1,0,9,4,z,z,z}
```

With ArrayPad we can add reference lists; for example, add a new list of values either left or right.

```
In[134]:= newVal={0,1,4,9}; (*Here we add them on the left side*)
ArrayPad[list4,{4,0},newVal]
Out[134]= {4,9,0,1,Blue,Soccer,1,0,9,4}
```

ArrayPad also has the functionality to remove elements from a list symmetrically using negative indices.

```
In[135]:= ArrayPad[list4,-1](*it deletes the first and last elemnts*)
Out[135]= {Soccer,1,0,9}
```

Note   With ArrayPad, addition and deletion is symmetric unless otherwise specified.

Structuring List

When we work with lists, in addition to the different forms of access to its content and content remover, we can come to present the case in which a discussion list is accommodated, sectioned, or restricted. Next, we will see several ways in which these tasks can be performed.

To sort a list into a specific order, use Sort followed by the sorting function.

```
In[136]:= Sort[{1,12,2,43,24,553,65,3},Greater]
Out[136]= {553,65,43,24,12,3,2,1}
```

Sort by default sorts values from less to greater, either numbers or text.

```
In[137]:= Sort[{"b","c","zz","sa","t","p"}]
Out[137]= {b,c,p,sa,t,zz}
```

To reverse a list, use the command Reverse.

```
In[138]:= Reverse[{1,12,2,43,24,553,65,3}]
Out[138]= {3,65,553,24,43,2,12,1}
```

To create a nested list in addition to that previously seen, you can generate partitions to a flat list by rearranging the elements of the list. For example, we will create partitions of a list to subdivide the list into pairs.

```
In[139]:= Partition[{1,12,2,43,24,553,65,3},2]
Out[139]= {{1,12},{2,43},{24,553},{65,3}}
```

We can choose a partition with successive elements inclused.

```
In[140]:= Partition[{1,12,2,43,24,553},3,1]
Out[140]= {{1,12,2},{12,2,43},{2,43,24},{43,24,553}}
```

Depending on how we want our nested list, we can add an offset to the partition. For example, a partition in two with an offset of four.

```
In[141]:= Partition[{"b","c","zz","sa","t","p"},2,4]
Out[142]= {{b,c},{t,p}}
```

To return to a flat list, the function used is Flatten.

```
In[143]:= Flatten[{{1,12},{2,43},{24,553},{65,3}}]
Out[143]= {1,12,2,43,24,553,65,3}
```

Depending on the depth of the list, we can decide how deep the Flatten should be.

```
In[144]:= Flatten[{{{{1},1},1},1},1] (* here we flatten a list with a level
1 depth.*)
Out[144]= {{{1},1},1,1}
```

When we have a list or an array we can reshape data into a specific rectangular array with ArrayReshape. For example, create an array of 3 by 3.

```
In[145]:= ArrayReshape[{1,12,2,43,24,553,65,3},{3,3}]
Out[145]= {{1,12,2},{43,24,553},{65,3,0}}
```

If you pay attention, you can see that elements that complete the array form are zeros. We can see this in the next example using ArrayShape to create an array of 2 by 2 from one element in the list.

```
In[146]:= ArrayReshape[{6},{2,2}]
Out[146]= {{6,0},{0,0}}
```

In the case when dealing with a nested list, SortBy is also used, but instead of a sorting function, a built-in function is used. For example, order a list by the result of their approximate value.

```
In[147]:= SortBy[{1,4,553,12.52,4.3,24,7/11},N]
```

$$Out[147]= \{\frac{7}{11},1,4,4.3,12.52,24,553\}$$

Criteria Selection

Particular values of a list can be selected with certain conditions; conditions can be applied to lists by using the command Select. The function selects the elements of the list that are true to the criteria established; the functions used for criteria can be order functions.

```
In[148]:= nmbrList=List[12,5,6,345,7,3,1,5];
Select[nmbrList,EvenQ] (* only the values that return True are selected, in
this case values that are even *)
Out[148]= {12,6}
```

Pick is also an alternative to Select.

```
In[149]:= Pick[nmbrList,PrimeQ @ nmbrList]
Out[149]= {5,7,3,5}
```

Pattern matching is used in the Wolfram Language to decree whether a criterion should be attributed to an expression. In the Wolfram Language, there are three types of patterns.

1. The underscore symbol (_) represents any expression in the Wolfram Language.

2. The double underscore symbol (__) represents a sequence of one or more expressions.

3. The triple underscore symbol (___) represents a sequence of zero or more expressions.

Every pattern has its built-in function name. One underscore is Blank, two underscores is BlankSequence, and three underscores is BlankNullSequence.

To better understand the following examples in the channels, we use the Cases function, which allows us to select data that corresponds to the pattern.

We have a list that consists of data pairs, and we write the selection pattern (_).

```
In[150]:= Cases[{{1,1},{1,2},{2,1},{2,2}},{_}]
Out[150]={}
```

As we can see, it does not choose any element, because it does not have the form of the pattern of the list—for example, the form {a, b}. Now if we change this shape, we will see that it selects all the elements that match the shape of the pattern.

```
In[151]:= Cases[{{1,1},{1,2},{2,1},{2,2}},{_,_}]
Out[151]= {{1,1},{1,2},{2,1},{2,2}}
```

The same result can be obtained if we use the double underscore.

```
In[152]:= Cases[{{1,1},{1,2},{2,1},{2,2}},{__}]
Out[152]= {{1,1},{1,2},{2,1},{2,2}}
```

In the following example we will see how we can select data from a list that contains numerical and categorical data. We use the RandomChoice function, which gives us a random selection from a list. In this case it is a random selection between the word Red or Blue. In the next chapter, we will see how this type of random functions works in the Wolfram Language.

```
In[153]:= Tbl=Table[{i,j,k,RandomChoice[{"Red","Blue"}]},{i,1,3},{j,1,3},
{k,1,3}]//TableForm
Out[153]//TableForm=
```

1 1 1	Red	1 2 1	Red	1 3 1	Blue
1 1 2	Red	1 2 2	Blue	1 3 2	Red
1 1 3	Blue	1 2 3	Blue	1 3 3	Blue
2 1 1	Blue	2 2 1	Red	2 3 1	Blue
2 1 2	Blue	2 2 2	Blue	2 3 2	Blue
2 1 3	Red	2 2 3	Red	2 3 3	Red
3 1 1	Blue	3 2 1	Blue	3 3 1	Red
3 1 2	Red	3 2 2	Red	3 3 2	Blue
3 1 3	Red	3 2 3	Blue	3 3 3	Blue

We can see that the numbers have on the right side the name Red or Blue. For example, we can use Cases to choose the values that belong to the Blue or Red category. Since this is a nested list of depth four, we have to specify the level at which Cases should search for patterns.

```
In[154]:= Cases[Tbl,{_,_,_,"Blue"},4]
Out[154]= {{1,1,3,Blue},{1,2,2,Blue},{1,2,3,Blue},{1,3,1,Blue},{1,3,3,Blue},
{2,1,1,Blue},{2,1,2,Blue},{2,2,2,Blue},{2,3,1,Blue},{2,3,2,Blue},{3,1,1,Blue},
{3,2,1,Blue},{3,2,3,Blue},{3,3,2,Blue},{3,3,3,Blue}}
```

Furthermore, the same result can be obtained using the double underscore.

```
In[155]:= Cases[Tbl,{__,"Blue"},4]
Out[155]= {{1,1,3,Blue},{1,2,2,Blue},{1,2,3,Blue},{1,3,1,Blue},{1,3,3,Blue}
,{2,1,1,Blue},{2,1,2,Blue},{2,2,2,Blue},{2,3,1,Blue},{2,3,2,Blue},{3,1,1,Bl
ue},{3,2,1,Blue},{3,2,3,Blue},{3,3,2,Blue},{3,3,3,Blue}}
```

We can even count how much of the Blue category we have.

```
In[156]:= Count[Tbl,{__,"Blue"},4]
Out[156]= 15
```

Count works in the next form, Count["list", pattern, level of spec].

Now that we know how the underscore symbol works, we can use the function Cases in checking conditions and filtering values. To attach a condition, use the next form (/; "condition"), when using the symbol /; followed by a rule or pattern, we are telling Mathematica that the next expression is a condition or pattern. In the next example, the x_ represents an arbitrary element x. In this case it represents the elements of the list and then the condition that x is greater than 5.

```
In[157]:= Cases[nmbrList,z_ /;z>5]
(*only the values greater than 5 are selected.*)
(*x can be replaced by any arbitrary symbol try using z_ and z > 5, the
result should be the same *)
Out[157]= {12,6,345,7}
```

As we saw in the previous example, what happens when we use _, means that the expression x_ must be applied to the condition > 5, since _ means any expression, which is the list.

Cases can be also used it select the data in which the condition is true of the established pattern or set of rules. In the next example, we are going to select data that are integers. The pattern objects are represented by an underscore or a rule of expression. For example,

```
In[158]:=mixList={1.,1.2,"4",\[Pi],{"5.2","Dog"}, 3,66,{Orange,Red}};
Cases[mixList,_Integer](*We now select the number that are integers*)
Out[158]= {3,66}
```

As you might notice, the use of the underscore can be applied to patterns that check the head of an expression, which is Integer. Cases compares each element to see if they are integers.

As for conditional matching, if the blanks of a pattern are accompanied by a question mark (?) and then the function test, the output is a Boolean value.

```
In[159]:= MatchQ[mixList,_?ListQ](*we test if mixlist has a head of List*)
Out[159]= True
```

We can select the level of specification with Cases. In the next example, we select the cases that are a string; we write two as a level of specification, because mixList is a nested list, with two sublists.

```
In[160]:= Cases[mixList,_?StringQ,2]
Out[160]= {4,5.2,Dog}
```

We can include several patterns, with alternatives. To test different alternatives, we just place a (|) between patterns, so it resembles the form "pattern1" | "pattern2" |"pattern3 "|...

```
In[161]:= Cases[mixList, _?NumberQ| _?String] (*We select the numbers and
the strings*)
Out[161]= {1.,1.2,3,66}
```

CHAPTER 3

Working with Data and Datasets

In this chapter, we will review the basics on how to apply functions to a list using Map and Apply; how to define user functions, which can be used throughout a notebook; and how to write code in one of the powerful syntax used in the Wolfram Language called pure functions. Naturally, we will pass, to the associations, how to associate keys with values and understand that they are fundamental for the correct construction of datasets in the Wolfram Language. We conclude with a final overview on how associations are abstract constructions of hierarchical data.

Operations with Lists

Now we will see how to perform operations on a list and between lists, this is important since, for the most part, results in Mathematica can be treated as lists. We will see how to perform arithmetic operations, addition, subtraction, multiplication, division, and scalar multiplication. We will detail how to apply functions to a list with the use of Map and Apply since these tools are useful when dealing with linear lists and nested lists because they allow us to specify the depth level of application of a function. We will see the approach of how to make user-defined functions, syntax, and term grouping; how to receive groups; and how to apply the function like any other. We will also review an important concept of the Wolfram Language, which is pure functions, since these are very important to be able to carry out powerful tasks and activities and write code in a compact way.

J. Villalobos Alva, *Beginning Mathematica and Wolfram for Data Science*,
https://doi.org/10.1007/978-1-4842-6594-9_3

Arithmetic Operations to a List

In this section we will see how lists support different arithmetic operations between numbers and between lists. We can perform basic arithmetic operations like addition, subtraction, multiplication, and division with the lists.

Addition and Subtraction

```
In[1]:= List[1,2,3,4,5,6]+1
Out[1]= {2,3,4,5,6,7}
```

```
In[2]:= List[1,2,3,4,5,6]-5
Out[2]= {-4,-3,-2,-1,0,1}
```

Division and multiplication

$$In[3]:= \text{List}[1,2,3,4,5,6] \ / \ \pi$$

$$Out[3]= \left\{ \frac{1}{\pi}, \frac{2}{\pi}, \frac{3}{\pi}, \frac{4}{\pi}, \frac{5}{\pi}, \frac{6}{\pi} \right\}$$

Perform scalar multiplication:

```
In[4]:= List[1,2,3,4,5,6]*2
Out[4]= {2,4,6,8,10,12}
```

Exponentiation

```
In[5]:= List[1,2,3,4,5,6]^3
Out[5]= {1,8,27,64,125,216}
```

Lists can also support basic arithmetic operations between lists.

```
In[6]:= List[1,2,4,5]-List[2,3,5,6]
Out[6]= {-1,-1,-1,-1}
```

We can also use mathematical notation to perform operations.

$$In[7]:= \frac{\left\{ "Dog",2 \right\}}{\{2,1\}}$$

$$Out[7]= \left\{ \frac{Dog}{2}, 2 \right\}$$

To perform computations between lists, the length of the lists must be the same; otherwise Mathematica will return an error specifying that lists do not have the same dimensions, like in the next example.

```
In[8]:= {1,3,-1}+{-1}
During evaluation of In[8]:= Thread::tdlen: Objects of unequal length in
{1,3,-1}+{-1} cannot be combined.
Out[8]= {-1}+{1,3,-1}
```

Joining a list

To join one list with another—that is, to join the two lists—there is the Union command, which joins the elements of the lists and shows it as a new list.

```
In[9]:= Union[List["1","v","c"],{13,4,32},List["adfs",3,1,"no"]]
Out[9]= {1,3,4,13,32,1,adfs,c,no,v}
```

In addition to the Union command, there is the Intersection command, which has a function analogous to what it represents in set theory. This command allows us to observe the common elements in the list or lists.

```
In[10]:= Intersection[{7,4,6,8,4,7,32,2},{123,34,6,8,5445,8}]
Out[10]= {6,8}
```

As seen the lists only have in common the numbers 6 and 8.

Applying Functions to a List

Functions can be applied to concise and automated to a list. The most used functions are Map and Apply. A short notation is to use the symbol @ instead of the square brackets [ ]; f@ "expr" is the equivalent to f[expr].

```
In[11]:= Max @ {1,245.2,2,5,3,5,6.0,35.3}
Out[11]= 245.2
```

Map has the following form, Map[f, "expr"]; another way of showing it is with the shorthand notation using the symbol @. f /@ "expr" and Map[f, "expr"] are equivalent. Nested lists are included too.

```
In[12]:= Factorial/@List[1,2,3,4,5,6]
Out[12]= {1,2,6,24,120,720}
```

Map can be applied to nested lists.

```
In[13]:= Map[Sqrt,{{1,2},{3,4}}]
Out[13]= {{1,Sqrt[2]},{Sqrt[3],2}}
```

With the use of the Map, the function is applied to each element of the list. Map can also work with nested lists, as in the previous example. In the next example, we will create a list of 10 elements with Table. Those elements will be a random number between 0 and 1, and then we will map a function to convert them to string expressions.

```
In[14]:= Data=Range[RandomReal[{0,1}],10];(*List *)
ToString/@Data (* mapping a to convert to string*)
Head/@ % (*Checking the data type of every element *)
Out[14]= {0.539347,1.53935,2.53935,3.53935,4.53935,5.53935,6.53935,7.53935,
8.53935,9.53935}
Out[14]= {String,String,String,String,String,String,String,String,String,
String}
```

We will see how to apply a function to a list with some additional functions. Apply has the form Apply [f, "expr"] and the shorthand notation is f @@ "expr".

```
In[15]:= Apply[Plus,Data](*It gives the sum of the elements of Data*)
Out[15]= 50.3935
```

```
In[16]:= Plus@@Data
Out[16]= 50.3935
```

Also, commands can be applied to a list in the same line of code. This can be helpful when dealing with large lists; for example, if we want to know whether the element satisfies a condition, then instead of going through each value, elements can be gathered into a list and then tested for the specified condition.

```
In[17]:= Primelist=Range[100];Map[PrimeQ,Primelist]
Out[17]= {False,True,True,False,True,False,True,False,False,False,True,
False,True, False,False,False,True,False,True,False,False,False,True,False,
False,False,False,False,True,False,True,False,False,False,False,False,True,
False,False,False,True,False,True,False,False,False,True,False,False,False,
False,False,True,False,False,False,False,False,True,False,True,False,False,
```

False,False,False,True,False,False,False,True,False,True,False,False,False,
False,False,True,False,False,False,True,False,False,False,False,False,True,
False,False,False,False,False,False,False,True,False,False,False}

In the latter example, we create a list from 1 to 100 and then test which of the numbers satisfies the condition of being a prime number with function PrimeQ. Other functions can be used to test different conditions with numbers and with strings. Also, a more specific function for testing logical relations in a list can be used (MemberQ, SubsetQ).

Defining Own Functions

User functions can be written to perform repetitive tasks and to reduce the size of a program. Segmenting the code into functions allows you to create pieces of code that perform a certain task. Functions can receive data from outside when called through parameters and return a fixed result.

A function can be defined with the set or set delayed symbol, but remember using a set symbol will assign the result to the definition. To define a function, we first write the name or symbol, followed by the reference variable and an underscore. Just as we saw with the use of cases, the underscore tells Mathematica that we are dealing with a dummy variable. As a warning, defined functions cannot have space between letters. Functions can also receive more than one argument.

```
In[18]:= MyF[z_]:=12+2+z;MyF2[x_,z_]:=z/x
```

Now we can call our function with different z values.

```
In[19]:= List[MyF[1],MyF[324],MyF[5432],MyF2[154,1],MyF2[14,4],MyF2[6,9]]
```

$$Out[19]= \left\{15,338,5446,\frac{1}{154},\frac{2}{7},\frac{3}{2}\right\}$$

Also, another way to write functions is to write compound functions. This concept is similar to compound expressions; expression of different classes are written within the definition. Each computation can or cannot be ended with a semicolon. The following example will show the concept.

```
In[20]:= StatsFun[myList_]:=
{
Max@myList,
Min@myList,
Mean@myList,
Median@myList,
Quantile@@{myList,1}(* 25 percent *)(* to write a function with multiple
arguments with shorthand notation use curly braces*)
}
```

And we can send a list as an argument.

```
In[21]:= myList=Table[m-2,{m,-2,10}];
StatsFun[myList]
Out[21]= {8,-4,2,2,8}
```

We can have multiple operations within a function, with the option to create conditions for the arguments to meet. To write a condition, we use the symbols dash and semicolon (/;) when the condition is true the function is evaluated; otherwise, if the condition is not true the function will not evaluate. Compound functions need to be grouped; otherwise Mathematica will treat them as though they are outside the body of the whole function.

In the next example, we will create a function that tells us if an arbitrary string is a palindrome, which is when the word is the same written backward.

```
In[22]:= PalindromeWord[string_/; StringQ @ string==True]:= (*we can check
if the input is really a string*)

(
ReverseWord=StringJoin[Reverse[Characters[string]]]; (*here we separate the
characters, reverse the list and join them into a string*)

ReverseWord ==string (* then we test if the word is a palindrome, the
output of the whole function will be True or False*)
)
```

We test our new function.

```
In[23]:= PalindromeWord/@{"hello","room","jhon","kayak","civic","radar"}
Out[23]= {False,False,False,True,True,True}
```

When we have a local assignment on a compound function or functions, the symbols used will still be assigned, so if the symbol(s) are called outside the function, it can cause coding errors. One thing to consider is when the function is no longer used, we can clear our function and local symbols. Clearing only the function name will not remove local assignments. Another solution is to declare variables inside a module, since the variables will only be locally treated. This can be seen in the following form.

```
In[24]:=
MyFunction[a0_,b0_]:=
Module[{m=a0,n=b0},(*local variables*)
m+n (*body of the module*)
](*end of module*)

In[25]:= Clear[MyF,MyF2,StatsFun,PalindromeWord,ReverseWord] (*To remove
tag names of the functions and local symbols *)
```

Pure Functions

Pure functions are a very powerful functionality of the Wolfram Language. It is possible to execute a function without referencing a name and have to explicitly assign a name function to an operation we want to execute. The arguments are tagged with a hashtag (#). To tag a specific argument, the symbol is followed up with a number; so, #1, #2, #3, ..., means argument one, two, three, ... Then an ampersand (&) is used at the end of the definition to mark the reference that will be used by the hashtag. Pure functions can be constructed with Function and with the shorthand notation of the hashtag and ampersand.

```
In[26]:=
Function[#^-1][z] == #^-1&[z]
#^-1&[z] (* both expression mean 1/z *)
Out[26]= True
Out[27]= 1/z
```

Some examples of pure functions.

```
In[28]:= {#^-1&[77],#1+#2-#3&[x,y,z] (* we can imagine that #1,#2,#3 are
the 1st, 2nd and 3rd variables*),
Power[E,#]&[3]}
Out[28]= {1/77,x+y-z, e³}
```

We can use pure functions along with Map and Apply, in the sense to pass each argument of a list to a specific function. The # represents each element of the list and the & represents that # is filled and tested for the elements of the list.

```
In[29]:= N[#]&/@ {1,1,1,12,3,1}
Sqrt[#]&/@{-1,2,4,16}
Out[29]= {1.,1.,1.,12.,3.,1.}
Out[30]= {I,Sqrt[2],2,4}
```

Code can be written in a more compact form with the use of Apply and pure functions, as shown in the next example; we can select the numbers that are bigger than 10.

```
In[31]:= Select@@{{1,22,41,7,62,21},#>10&}
Out[31]= {22,41,62,21}
```

Indexed Tables

To provide a quick way to observe and manage a group of related data, we can create and display results in the form of tables. This leads us to see how to create tables in the Wolfram Language, such as giving titles to columns and names to rows. We will expose a series of examples that will help you learn the essentials to use the tables so that you can present your data properly.

Tables with the Wolfram Language

Tables are created with nested lists, and those lists are portrayed with the function TableForm.

```
In[32]:= table1={{"Dog","Wolf"},{"Cat","Leopard"},{"Pigeon","Shark"}};
TableForm[table1]
Out[32]//TableForm=
```

```
Dog        Wolf
Cat        Leopard
Pigeon     Shark
```

The format of TableForm is ["list", options]. Formatting options lets you justify the columns of tables in three ways: left, center, and right. In the next example, the contents of the table are centered.

```
In[33]:= TableForm[table1,TableAlignments → Right]
Out[33]//TableForm=

      Dog    Wolf
      Cat Leopard
   Pigeon    Shark
```

Titles can be added with the option command TableHeadings and by specifying whether both the rows and columnlabels or just one of them will be exposed. Choosing the option Automatic gives index labels to the rows and columns. Remember to write strings between the apostrophes or to use ToString.

```
In[34]:= TableForm[table1,TableHeadings->{{"Row 1","Row 2","Row 3"},
{"Column 1","Column 2","Column 3"}}]
Out[34]//TableForm=
```

	Column 1	Column 2
Row 1	Dog	Wolf
Row 2	Cat	Leopard
Row 3	Pigeon	Shark

Labeled rows and columns can be customized to desired names.

```
In[35]:= Colname={"Domestic Animals","Wild Animals"};
Rowname={"Animal 1","Animal 2","Animal 3"};
TableForm[table1,TableHeadings → {Rowname,
Colname}]
Out[35]//TableForm=
```

	Domestic Animals	Wild Animals
Animal 1	Dog	Wolf
Animal 2	Cat	Leopard
Animal 3	Pigeon	Shark

The same concept applies to label just columns or rows by typing None on rows or columns option.

```
In[36]:= TableForm[table1,TableHeadings → {None, {"Domestic Animals","Wild
Animals"}}]
Out[36]//TableForm=
```

Domestic Animals	Wild Animals
Dog	Wolf
Cat	Leopard
Pigeon	Shark

Automated forms of tables can be created with the use of Table and Range. By applying the Automatic option in the TableHeadings, we can create indexed labels for our data.

```
In[37]:= TabData={Table[i,{i,10}],Table[5^i,{i,10}]};
TableForm[TabData,TableHeadings → Automatic]
Out[37]//TableForm=
```

	1	2	3	4	5	6	7
1	1	2	3	4	5	6	7
2	5	25	125	625	3125	15625	78125

For exhibit reasons, a table can be transposed too.

```
In[38]:= TableForm[Transpose[TabData],TableHeadings → Automatic]
Out[38]//TableForm=
```

	1	2
1	1	5
2	2	25
3	3	125
4	4	625
5	5	3125
6	6	15625
7	7	78125

Another useful tool is Grid. With Grid we can display a list or a nested list in tabular format. Just like with TableForm, Grid can also be customized to exhibit data more properly.

Note   Grid works with any kind of expression.

```
In[39]:= TabData2 = Table[{i,Exp[i],N @ Exp[i]},{i,7}];
Grid[TabData2]
Out[39]=
```

1	e	2.71828
2	e^2	7.38906
3	e^3	20.0855
4	e^4	54.5982
5	e^5	148.413
6	e^6	403.429
7	e^7	1096.63

To add headers, we need to insert them in the original list as strings and in position 1.

```
In[40]:= Grid[Insert[TabData2,{"i","Exp i","Numeric approx."},1]]
Out[40]=
```

i	Exp i	Numeric approx.
1	e	2.71828
2	e^2	7.38906
3	e^3	20.0855
4	e^4	54.5982
5	e^5	148.413
6	e^6	403.429
7	e^7	1096.63

We can add dividers and spacers too. With Dividers and Spacing we can choose to divide or space the y and x axes.

```
In[41]:=Grid[Insert[TabData2,{"i","Exp i","Numeric approx."}, 1],Dividers
→ {All, False},Spacings → {1, 1} ]
Out[41]=
```

i	Exp i	Numeric approx.
1	e	2.71828
2	e^2	7.38906
3	e^3	20.0855
4	e^4	54.5982
5	e^5	148.413
6	e^6	403.429
7	e^7	1096.63

Background can be added with the Background option. With this option specific parts of the table or column table can be colored.

```
In[42]:=Grid[Insert[TabData2,{"i","Exp i","Numeric approx."},
1],Dividers → {All, False},Spacings → {Automatic, 0},Background →
{{LightYellow,None,LightBlue}}]
Out[42]=
```

i	Exp i	Numeric approx.
1	e	2.71828
2	e^2	7.38906
3	e^3	20.0855
4	e^4	54.5982
5	e^5	148.413
6	e^6	403.429
7	e^7	1096.63

Associations

Associations are fundamental in the development of the Wolfram Language; associations are used for indexing lists or other expressions and creating more complex data structures. Associations are a more structured construct that allow us to provide a process for creating pairs of key and value. Associations are useful for looking at a particular value given the key. Later we will see that they are important for handling data sets in the Wolfram Language.

Associations are of the form Association["key_1" → val_1, key_2 →val_2 ...] or < | "key_1"→ "val_1", "key_2" → "val_2" ... | >; they associate a key to a value. Keys and values can be any expression. To construct an association, the Association command is used or we can use symbolic entry <| --- |>.

```
In[43]:=
Associt=<|1 → "a",2 → "b",3 → "c"|> (* is the same as Association[a → "a",
b → "b",c → "c"] *)
Associt2= Association[ dog → "23","score" → π * π, 2*2 → Sin[23 Degree]]
Out[43]= <|1 → a,2 → b,3 → c|>
Out[44]= <|dog → 23,score → π²,4 → Sin[23 °] |>
```

Entries in an association are ordered so data can be accessed based on the key of the value or by the elements of the association, like with lists. The position is associated with the values, not the key.

```
In[45]:= Associt[1](*this is key 1 *)
Associt2[[2]] (*this is position of key 2, which is π² *)
Out[45]= a
Out[46]= π²
```

As seen in the latter example, the position is associated with the values, not the key. So, if we want to show parts of the association, we can use the semicolon.

```
In[47]:= Associt[[1;;2]]
Associt2[[2;;2]]
Out[47]= <|1 → a,2 → b|>
Out[48]= <|score → π² |>
```

Values and keys can be extracted with the command Keys and Values.

```
In[49]:= Keys@Associt2
Values@Associt2
Out[49]= {dog,score,4}
Out[50]= {23, π²,Sin[23 °]}
```

If we asked for a key without proper reference, we will get an error.

```
In[51]:= Associt["a"](* there is no "a" key in the association, thus the
error*)
Out[51]= Missing[KeyAbsent,a]
```

Association can also be associations. As seen in the next example, we can associate associations, thus producing an association of associations. This concept is basic for understanding how a dataset works in the Wolfram Language.

```
In[52]:= Association[Associt,Associt2]
Out[52]= <|1 → a,2 → b,3 → c,dog → 23,score → π²,4 → Sin[23 ° ]|>
```

We can also make different associations with lists using AssociationThread. The keys correspond to the first argument and the values to the second. This is achieved with the command AssociationThread. AssociationThread threads a list of keys to a list of values like the next form: < | {"key_1", "key_2", "key_3" ...} → {"val_1", "val_2", "val_3" ... | >. The latter form can be seen as a list of keys marking to a list of values. The command can be used to associate a list with a list when we have defined our lists of keys and values. We can also create a list of associations in a way that keys can be read as a row and a column.

```
In[53]:= AssociationThread[{"class","age","gender","survived"},{"Economy",
29,"female",True}]
Out[53]= <|class → Economy,age → 29,gender → female, survived→ True|>
```

We can construct our list of keys and values.

```
In[54]:= keys={"class","age","gender","boarded"};
values={"Economy",29,"female",True};
AssociationThread@@{keys,values}
Out[54]= <|class → Economy,age → 29,gender → female,boarded → True|>
```

More complex structures can be done with associations. For example, the next association creates a data structure based on the information about a sports car, with the Model name, Engine, Power, Torque, Acceleration, and Top speed.

```
In[55]:= Association@
{
"Model name" → "Koenigsegg CCX",
"Engine" → "Twin supercharged V8",
```

```
"Power" → 806 "hp",
"Torque" → 5550 "rpm",
"Acceleration 0-100 km/h" → 3.2 "sec",
"Top speed" → 395 "Km/h"
}
Out[55]= <|Model name → Koenigsegg CCX,Engine → Twin supercharged
V8,Power → 806 hp,Torque → 5550 rpm,Acceleration 0-100 km/h →
3.2 sec,Top speed → 395 Km/h|>
```

We can see how labels and their elements are created in a grouped way. In addition to that, it is shown how the curly braces mark how the key/value pair can be arranged by each row.

Dataset Format

As we have seen through later sections, associations are an essential part of making structure forms of data. As we will see, datasets in the Wolfram Language offer a way to organize and exhibit hierarchical data by providing a method for accessing data inside a dataset. We will see examples throughout this section, including how to convert lists, nested lists, and associations to a dataset. In addition, we will cover topics on how to add values, access values of a dataset, dropping and deleting values, mapping functions over a dataset, dealing with duplicate data, and applying functions by row or column.

Constructing Datasets

Datasets are for constructing hierarchical data frameworks, where lists, associations, and nested lists have an order. Datasets are useful to exhibit large data in an accessible structured format. With datasets, we can show enclosed structures in a sharp format with row headers, column headers, and numbered elements. Having the data as a dataset allows us to look at the data in multiple ways.

Datasets can be constructed in four forms.

1. A list of lists; a table with no denomination in rows and columns

2. A list of associations, a table with labeled columns; a table with repeated keys and different or same values

3. An association of lists, a table with labeled rows; a table with different keys and different or same values

4. Association of associations; a table with labeled rows and columns

Datasets can be created manually or by creating associations first. First, before using the datasets, we must make sure that "Dynamic Update" is enabled; otherwise the Mathematica kernel will not display the dataset correctly and we will only see an overlapping cell in case of evaluating the expression. So, if we hover the mouse pointer to the data set object content it will show us an advertisement that the dynamic content cannot be shown. To activate it, go to the menu bar, then Evaluation, and select Dynamic Updating Enabled.

The most common form to create a new dataset is from a list of lists. Using the function Dataset, we create a list within the curly braces {}. Each brace represents the parts of our table. Figure 3-1 shows the output of the Dataset function.

```
In[56]:= Dataset@
{
{"Jhon",23,"male","Portugal"},
{"Mary",30,"female","USA"},
{"Peter",33,"male","France"},
{"Julia",53,"female","Netherlands"},
{"Andrea",45,"female","Brazil"},
{"Jeff",24,"male","Mexico"}
}
Out[56]=
```

Jhon	23	male	Portugal
Mary	30	female	USA
Peter	33	male	France
Julia	53	female	Netherlands
Andrea	45	female	Brazil
Jeff	24	male	Mexico

Figure 3-1. *Dataset object created from the input code*

By hovering the mouse cursor in the elements of the dataset, you can see the position in the lower-left corner. The name France corresponds to position row 3 and column 4. The notation of a datasets is first rows, then columns. In the case we have label column, rows or both, instead of the numbers, we will see the column name and row name.

Constructing a data set by of list of associations is performed by creating associations first with repeated keys and then enclosing by a list. First, we create our associations; the repeated keys specify each column header. The values represent the contents of the columns. Datasets have a head expression of Dataset.

```
In[57]:= Dataset@
{
<|"Name" → "Jhon", "Age" → 23, "Gender" → "male", "Country" → "Portugal" |>,
<|"Name" → "Mary", "Age" → 30, "Gender" → "female", "Country" → "USA" |>,
<|"Name" → "Peter", "Age" → 33, "Gender" → "male", "Country" → "France" |>,
<|"Name" → "Julia", "Age" → 53, "Gender" → "female", "Country" →
"Netherlands"|>,
<|"Name" → "Andrea", "Age" → 45, "Gender" → "female", "Country" →
"Brazil |>,
<|"Name" → "Jeff", "Age" → 24, "Gender" → "male", "Country" → "Mexico" |>
}
(*Head @ % *)

Out[57]=
```

Name	Age	Gender	Country
Jhon	23	male	Portugal
Mary	30	female	USA
Peter	33	male	France
Julia	53	female	Netherlands
Andrea	45	female	Brazil
Jeff	24	male	Mexico

Figure 3-2. *Dataset with column headers*

As can be seen in Figure 3-2, Mathematica recognizes that Name, Age, Gender, and Country are column headers, which is why the color of the box is different. When passing the cursor over the column labels, they will be highlighted in blue, thus making it possible to click the name of the label, and then it will produce only the selected label and not the whole dataset, as seen in Figure 3-3.

Figure 3-3. *Column name selected in the dataset*

When this happens the name of the column will also appear; to return to the whole dataset, hit the spreadsheet icon ▦ on the higher left corner or the name All. This type of layout is practical when we are dealing with a big set of rows and columns, and we want to focus only on a few sections of our data set.

An Association of lists, in this case, the keys represent the label of the rows and the values are the list of the elements of the rows; then we associate the whole block. The next block of code generates an association of a list.

Note   The same is applied here. Whenever you click the name of the row, it will only display that row.

```
In[58]:= Dataset@
<|
"Subject A" → {"Jhon", 23, "male", "Portugal"},
"Subject B" → {"Mary", 30, "female", "USA"},
"Subject C" → {"Peter", 33, "male", "France"},
"Subject D" → {"Julia", 53, "female", "Netherlands"},
"Subject E" → {"Andrea", 45, "female", "Brazil"},
```

```
"Subject F" → {"Jeff", 24, "male", "Mexico"}
|>
Out[58]=
```

Subject A	Jhon	23	male	Portugal
Subject B	Mary	30	female	USA
Subject C	Peter	33	male	France
Subject D	Julia	53	female	Netherlands
Subject E	Andrea	45	female	Brazil
Subject F	Jeff	24	male	Mexico

Figure 3-4. *Dataset with labeled rows*

As can be seen in Figure 3-4, the rows are now labeled. Just like the previous examples, row labels are recognized and displayed in the color box. When selecting the label of the row, it will display only that row, as shown in Figure 3-5.

⊞ › Subject E			
Andrea	45	female	Brazil

Figure 3-5. *Subject E row selected*

Association of associations, in this form, the repeated keys of the association of associations are the column labels and the values of the content of the dataset. In the second association, the keys are the labels of the rows, and the first associations are the values of the second association. The next example clarifies this.

```
In[59]:= Dataset@
<|
"Subject A" → <|"Name" → "Jhon", "Age" → 23, "Gender" → "male",
"Country" → "Portugal"|>,
"Subject B" → <|"Name" → "Mary", "Age" → 30, "Gender" → "female",
"Country" → "USA"|>,
"Subject C" → <|"Name" → "Peter", "Age" → 33, "Gender" → "male",
"Country" → "France"|>,
```

```
"Subject D" → <|"Name" → "Julia", "Age" → 53, "Gender" → "female",
"Country" → "Netherlands"|>,
"Subject E" → <|"Name" → "Andrea", "Age" → 45, "Gender" → "female",
"Country" → "Brazil" |>,
"Subject F" → <|"Name" → "Jeff", "Age" → 24, "Gender" → "male",
"Country" → "Mexico"|>
|>
Out[59]=
```

	Name	Age	Gender	Country
Subject A	Jhon	23	male	Portugal
Subject B	Mary	30	female	USA
Subject C	Peter	33	male	France
Subject D	Julia	53	female	Netherlands
Subject E	Andrea	45	female	Brazil
Subject F	Jeff	24	male	Mexico

Figure 3-6. *Dataset with names in rows and columns*

As can be seen in Figure 3-6, the rows and columns are now labeled. Just like the previous examples, the column and row labels are recognized and displayed in the color box. When selecting the label of the row or a column it will display only that row or column, as seen in Figure 3-7.

> ⊞ › Subject F
>
> | Name | Jeff |
> |---|---|
> | Age | 24 |
> | Gender | male |
> | Country | Mexico |

Figure 3-7. *Only a row selected*

If we select only a particular value, then that value is solely displayed. Figure 3-8 shows its form.

Figure 3-8. *Name for subject F*

Creating a dataset from associations of associations are best for compact dataset purposes because sometimes it can get messy trying to extract values and keys. However, the best approach is the one that best works for you.

Accessing Data in a Dataset

Mathematica gives each element a unique index, so if we are interested in selecting data from our dataset, we then assign a symbol to the dataset and proceed to specify each output in the next form. The first position and second position of the arguments represent row and column [n[th] row, m[th] column]. So, to extract data based on a column name or a set of columns, we enclose the columns in brackets. We can also use the double bracket notation. In the case that only one argument is received, it will only be the rows.

First let us create the data set.

```
In[60]:= Dst=Dataset@
{
<|"Name" → "Jhon", "Age" → 23, "Gender" → "male", "Country" →
"Portugal"|>,
<|"Name" → "Mary", "Age" → 30, "Gender" → "female", "Country" →
"USA"|>,
<|"Name" → "Peter", "Age" → 33, "Gender" → "male", "Country" →
"France"|>,
<|"Name" → "Julia", "Age" → 53, "Gender" → "female", "Country" →
"Netherlands"|>,
<|"Name" → "Andrea", "Age" → 45, "Gender" → "female", "Country" →
"Brazil"|>,
<|"Name" → "Jeff", "Age" → 24, "Gender" → "male", "Country" →
"Mexico"|>
};
```

The notation [[ ]] works the same as the special character for double brackets (⟦ ⟧) Also, we can select data with the specific keys of the value.

```
In[61]:= Dst[[1,2]](* This is for row 1, column 2*)
Dst[1](*row 1*)
Out[61]= 23
Out[62]=
```

Name	Jhon
Age	23
Gender	male
Country	Portugal

Figure 3-9. *Row 1 for Dst*

As shown in Figure 3-9, let´s see Figure 3-10.

```
In[63]:= Dst[1;;3](*to manipulate data of the column try Dst[1;;3, 1;;3] *)
Out[63]=
```

Name	Age	Gender	Country
Jhon	23	male	Portugal
Mary	30	female	USA
Peter	33	male	France

Figure 3-10. *Values from columns 1 to 3*

In this case, we selected data from positions 1 to 3—that is, from John to Peter. The same is applied for columns. The output is shown in Figure 3-10.

We can also show specific columns and maintain all the fixed rows with their keys. The same process is applied when having a label in each row. By typing All, we mean all elements of the column or the row. The output is shown in Figure 3-11.

```
In[64]:= Dst[All,{"Name","Age"}] (*If more than 1 column label is added
then enclosed the labels by curly braces. *)
Out[64]=
```

Name	Age
Jhon	23
Mary	30
Peter	33
Julia	53
Andrea	45
Jeff	24

Figure 3-11. *Values for column name and age*

As an alternative, we can extract a column or a row as a list in order to manipulate them better in the Wolfram Language. To do that we need to use the function Normal and Values. Remember that we are dealing with associations, so if we want the values then we use the command Values and then Normal to convert it to a normal expression.

```
In[65]:= Normal@Values@Dst[All,{"Name","Age"}](*values of the name and age columns*)
Out[65]= {{Jhon,23},{Mary,30},{Peter,33},{Julia,53},{Andrea,45},{Jeff,24}}
```

For the rows, it is the same idea: if the rows have a label, then we can use the label(s) of the rows.

```
In[66]:= Normal@ Values@Dst[[1,All]]
Out[66]= {Jhon,23,male,Portugal}
```

The result is the same if we first do Normal and then Values.

```
In[67]:= Values@Normal@Dst[[1,All]]
Out[67]= {Jhon,23,male,Portugal}
```

Another function that can be used is Query; this is a specialized function that works with datasets. Queries must be applied to the symbol of the dataset or directly to the dataset. Queries are helpful because it allows easy selectivity of the values; also we can extract rows or columns and get sole records.

```
In[68]:= Query[All,"Country"]@Dst
Query[3]@%
Out[68]=
```

Figure 3-12. *Country values*

```
Out[69]= France
```

As seen in Figure 3-12, we can extract columns and values with Query.

Another function that works more intuitively is Take, in which we can simply specify the symbol of our dataset and then how many rows and columns to display. Take comes handy when we are dealing with large data sets and we want to only view a specific part of our data.

```
In[70]:= Take[Dst,2] (*First 2 rows*)
(*Take [Dst, 3,3] First 3 rows and columns*)
Out[70]=
```

Name	Age	Gender	Country
Jhon	23	male	Portugal
Mary	30	female	USA

Figure 3-13. *First two rows of a dataset*

As seen in Figure 3-13, we can use Take as an alternative.

Adding Values

Now that we have examined how to access the elements of a data set, we can now proceed with how to add new values to our dataset. We can add rows with Append or Prepend, but remember AppendTo and PrependTo can be used too. However, they will assign the new result to the assigned variable. Append adds at the last and Prepend at the first.

To add a row, we would need to write the new row like we write the associations with repeated keys, calling our dataset and then the function, followed by the new row, as shown in Figure 3-14.

```
In[71]:= Dst[Append[<|"Name" → "Anya", "Age" → 19, "Gender" → "female",
"Country" → "Russia"|>]]
Out[71]=
```

Name	Age	Gender	Country
Jhon	23	male	Portugal
Mary	30	female	USA
Peter	33	male	France
Julia	53	female	Netherlands
Andrea	45	female	Brazil
Jeff	24	male	Mexico
Anya	19	female	Russia

Figure 3-14. *New row added at the end of the dataset*

To add a new row at the top of the dataset, try using the code, Prepend[Sst[Prepend[<|"Name" → "Anya", "Age" →19, "Gender" → "female", "Country" → "Russia"|>]]].

Adding a new column of only single values can be done by simply assigning a value to the side of the columns of the dataset with the key name, which is the column name. Figure 3-15 shows the new column added.

```
In[72]:= Dst[All,Prepend["ID number" → 1]]
Out[72]=
```

ID number	Name	Age	Gender	Country
1	Jhon	23	male	Portugal
1	Mary	30	female	USA
1	Peter	33	male	France
1	Julia	53	female	Netherlands
1	Andrea	45	female	Brazil
1	Jeff	24	male	Mexico

Figure 3-15. *ID column added*

To add a list of values as a column, we first need a list of values, and then we associate each value to the same key by using AssociationThread, creating an association of values for the repeated key. Then we proceed to create a dataset of the new association and joining objects, the entire dataset and the new one, with the function Join. Join can gather expressions of the same head.

```
In[73]:= Id={1,2,3,4,5,6};(* our list of values *)
ID=AssociationThread["ID" → #]&/@Id (* the process is threaded in the
list *)
Out[73]=
{<|ID → 1|>,<|ID → 2|>,<|ID → 3|>,<|ID → 4|>,<|ID → 5|>,<|ID → 6|>}
```

For the later block, each element needs to associate one by one, because AssociationThread suppresses repeated keys, so we would only have one association, and we need to have a repeated key marking to different values.

Now we proceed to create the new data set with the same key shown in Figure 3-16.

```
In[74]:= Dataset[ID]
Out[74]=
```

ID
1
2
3
4
5
6

Figure 3-16. *ID column dataset*

Finally we join the same objects; here we use Join with a level of specification of 2, because the new dataset is a sublist of depth 2. If we want to add the column of the left side, the new column goes first, and then the dataset; for the right side, it is the opposite. Figure 3-17 shows the output data set.

```
In[75]:= Join[%,Dst,2]
Out[75]=
```

ID	Name	Age	Gender	Country
1	Jhon	23	male	Portugal
2	Mary	30	female	USA
3	Peter	33	male	France
4	Julia	53	female	Netherlands
5	Andrea	45	female	Brazil
6	Jeff	24	male	Mexico

Figure 3-17. *ID column added*

In the previous cases, we worked with dataset from a list of associations; in the sense that we are working with tagged rows only or tagged rows and columns, the process of adding a row or column is preserved by adding the same structure to the dataset. So, adding a new row to an association of lists would take the form < | "key" → {elem, ... } | >; for columns, this would be the process of creating a dataset and joining them. In the case for a list of lists, adding a row would be the same approach but without a key. For the case of association of associations, to add a row would be <| "key" → < |"key 1" → "val 1", ... | > |>, and for columns, it would be the same as before, a key associated with a value. Nevertheless, there is no restriction on how data can be accommodated.

Finally, to change unique values, we select the item and give it the new content. In the case that we have labels on row and columns, the original form is still preserved, {"rows", "columns"}. So, if we want to replace the age of Jhon, we use the function ReplacePart by calling the symbol of the dataset and then specifying the column tag and then with the new value, which is 50. If we were working with only a row label or a column label the process would be the same but using the row or column label and then the number of position of the element.

```
In[76]:= ReplacePart[Dst, {1, "Age"} → 50](*Also using the index will
produce the same output, that would be {1,2} → 50 *)
Out[76]=
```

Name	Age	Gender	Country
Jhon	50	male	Portugal
Mary	30	female	USA
Peter	33	male	France
Julia	53	female	Netherlands
Andrea	45	female	Brazil
Jeff	24	male	Mexico

Figure 3-18. *Jhon age value changed to 50*

As seen in Figure 3-18 the new value is 50.

Dropping Values

We can eliminate the contents of a row or column without deleting the entire table structure. To accomplish this, we use the function Drop or Delete. When using Drop, we enclose the number of the row or column with { } to delete a unique row or column (Figure 3-19).

```
In[77]:= Drop[Dst,{1}](*in the instance we want to delete more than one
then we write m through n dropped {m,n} *)
Out[77]=
```

106

Name	Age	Gender	Country
Mary	30	female	USA
Peter	33	male	France
Julia	53	female	Netherlands
Andrea	45	female	Brazil
Jeff	24	male	Mexico

Figure 3-19. *Drop row 1*

As seen in Figure 3-19, we have dropped the first row. We can also drop rows and columns at the same time. Figure 3-20 shows the second row and last column dropped.

```
In[78]:=Drop[Dst,{2},{4}]
Out[78]=
```

Name	Age	Gender
Jhon	23	male
Peter	33	male
Julia	53	female
Andrea	45	female
Jeff	24	male

Figure 3-20. *New dataset after dropping row 2 and column 4*

Another way to do it, but by using the label of the key of a row or a column, is to use Delete, as shown in Figure 3-21.

```
In[79]:= Dst[All,Delete["Age"]] (*to delete a row use ["label of row",All] *)
Out[79]=
```

Name	Gender	Country
Jhon	male	Portugal
Mary	female	USA
Peter	male	France
Julia	female	Netherlands
Andrea	female	Brazil
Jeff	male	Mexico

Figure 3-21. *Age column deleted*

Filtering Values

Having the data as a dataset allows us to look at the data in multiple ways. Let's now work with the tagged dataset to better expose how filtering values work. For starters we will use the labeled dataset shown in Figure 3-22.

```
In[80]:= Clear[Dst];(* Let's clear the symbol "Dst" of previous
assignments *)
Dst=Dataset@
<|
"Subject A" → <|"Name" → "Jhon", "Age" → 23, "Gender" → "male",
"Country" → "Portugal"|>,
"Subject B" → <|"Name" → "Mary", "Age" → 30, "Gender" → "female",
"Country" → "USA"|>,
"Subject C" → <|"Name" → "Peter", "Age" → 33, "Gender" → "male",
"Country" → "France"|>,
"Subject D" → <|"Name" → "Julia", "Age" → 53, "Gender" → "female",
"Country" → "Netherlands"|>,
"Subject E" → <|"Name" → "Andrea", "Age" → 45, "Gender" → "female",
"Country" → "Brazil"|>,
"Subject F" → <|"Name" → "Jeff", "Age" → 24, "Gender" → "male",
"Country" → "Mexico"|>
|>
Out[80]=
```

	Name	Age	Gender	Country
Subject A	Jhon	23	male	Portugal
Subject B	Mary	30	female	USA
Subject C	Peter	33	male	France
Subject D	Julia	53	female	Netherlands
Subject E	Andrea	45	female	Brazil
Subject F	Jeff	24	male	Mexico

Figure 3-22. *Tagged dataset*

Just like with lists, we can create one or more filter conditions; for example, we can select the age that is greater than 30. And we will get a dataset object (Figure 3-23).

```
In[81]:= Cases[Dst[All,"Age"],x_/;x>30](*also we can select data that
matches exactly 30 with the == sign*)
Out[81]=
```

33
53
45

Figure 3-23. *Filtered data from the age column*

Figure 3-23 shows the filtered data. Data can be selected based on True or False results. For that, we can use the function Select.

```
In[82]:= Select[Dst[All,"Age"],EvenQ]
Out[82]=
```

Subject B	30
Subject F	24

Figure 3-24. *Selected subjects*

Figure 3-24 shows the selected subjects. The use of pure functions can be applied too. Remember that the #Age resembles the elements in the column Age, as shown in Figure 3-25.

```
In[83]:= Dst[Select[#Age>30&]]
Out[83]=
```

	Name	Age	Gender	Country
Subject C	Peter	33	male	France
Subject D	Julia	53	female	Netherlands
Subject E	Andrea	45	female	Brazil

Figure 3-25. *Selected values using pure function syntax*

Also, we can count the values of categorical data, as shown in Figure 3-26. This is helpful when we want to identify how many types of a class we have in our data. For example, we can count how many females and males are in the dataset.

```
In[84]:= Counts[Dst[All,"Gender"]] (*alternative form:Dst[Counts,"Gender"] *)
Out[84]=
```

male	3
female	3

Figure 3-26. *Count data for class male and female*

More complex groups can be made based on a class; for instance, we can group the dataset by gender, as shown in Figure 3-27.

```
In[85]:= Dst[GroupBy["Gender"],Counts,"Age"]
Out[85]=
```

male	23	1
	33	1
	24	1
female	30	1
	53	1
	45	1

Figure 3-27. *Data arranged by class and age*

As a good practice, let's clear symbols when they are not going to be used anymore.

```
In[86]:= Clear[Dst]
```

Applying Functions

Functions can be applied to our dataset. This class of functions in particular can be statistics, to know the dimension of our data or to transform data . Functions can be applied to single columns or to a unique element in the data structure.

First let's create a dataset that consists of 10 items, which columns will be the factorial of 1 to 10, a random real number from 1 to 0, and the natural logarithm from 1 to 10. Figure 3-28 shows the new dataset.

```
In[87]:= DataNumbr=Dataset@Table[<|"Factorial" → Factorial[i], "Random
number" → RandomReal[{0,1}], "Natural Logarithm" → Log[E,i]|>,{i,1,10}]
Out[87]=
```

Factorial	Random number	Natural Logarithm
1	0.983625	0
2	0.252122	0.693147
6	0.115411	1.09861
24	0.652804	1.38629
120	0.355895	1.60944
720	0.230432	1.79176
5040	0.945289	1.94591
40320	0.830881	2.07944
362880	0.197961	2.19722
3628800	0.202727	2.30259

Figure 3-28. *Numeric dataset*

And now we can compute basic operations to our data, like getting the mean of the factorials and random numbers, as shown in Figure 3-29.

```
In[88]:= DataNumbr[Mean,{"Factorial","Random number"}] //N
Out[88]=
```

Factorial	403791.
Random number	0.476715

Figure 3-29. *Mean for values in Factorial and Random number columns*

Parenthesis and composition of functions can also be used to relate operations applied to the data by using the @ *(composition) symbol. Figure 3-30 shows the data of Random numbers sorted from less to greater.

```
In[89]:= DataNumbr[All,"Random number"]@(Sort@*N)
Out[89]=
```

| 0.115411 | 0.197961 | 0.202727 | 0.230432 | 0.252122 |
| 0.355895 | 0.652804 | 0.830881 | 0.945289 | 0.983625 |

Figure 3-30. *Sorted data in canonical order*

We can apply different functions to our data. As can be seen in Figure 3-31, the dataset shows numbers in decimal form, because otherwise it would not fit in the square box.

```
In[90]:= DataNumbr[{Total,Max,Min},"Natural Logarithm"]
Out[90]=
```

| 15.1044 | 2.30259 | 0 |

Figure 3-31. *Total, Max, and Min value for Natural Logarithm column*

We can also apply our own created functions; let's use a previously constructed function.

```
In[91]:= StatsFun[myList_]:=
{
Max@myList,
Min@myList,
Mean@myList,
Median@myList,
Quantile@@{myList,1}(* 25 percent *) (* to write a function with multiple
arguments with shorthand notation use curly braces*)
}
```

Figure 3-32 shows the function we have created previously, applied to a column of the dataset.

```
In[92]:= DataNumbr[{StatsFun},"Natural Logarithm"]
Out[92]=
```

2.30259	0	1.51044	1.7006	2.30259

Figure 3-32. *StatsFun applied to the column Natural Logarithm*

Functions to restructure the dataset can be applied too, like Reverse, as shown in Figure 3-33.

```
In[93]:= DataNumbr[Reverse,All]
Out[93]=
```

Factorial	Random number	Natural Logarithm
1	0.991779	0
1.41421	0.502118	0.832555
2.44949	0.339723	1.04815
4.89898	0.807963	1.17741
10.9545	0.596569	1.26864
26.8328	0.480034	1.33857
70.993	0.97226	1.39496
200.798	0.911527	1.44203
602.395	0.444928	1.4823
1904.94	0.450252	1.51743

Figure 3-33. *Reversed elements of the dataset*

Map can also be used to apply functions like we saw with lists in the previous sections. In the next example we will map a function directly into our dataset, as we can see in Figure 3-34.

```
In[94]:= Map[Sqrt,DataNumbr]
Out[94]=
```

Factorial	Random number	Natural Logarithm
1	0.991779	0
1.41421	0.502118	0.832555
2.44949	0.339723	1.04815
4.89898	0.807963	1.17741
10.9545	0.596569	1.26864
26.8328	0.480034	1.33857
70.993	0.97226	1.39496
200.798	0.911527	1.44203
602.395	0.444928	1.4823
1904.94	0.450252	1.51743

Figure 3-34. *Square root function mapped in the dataset*

Transposition is an operation that consists of converting columns to rows and rows to columns and can sometimes help us to observe data in a different way. To obtain the transposition of the dataset, we use the Transpose function applied to the dataset. As we can see in Figure 3-35, all columns are now rows and displayed in a compact way because it is a large row.

```
In[95]:= DataNumbr//Transpose
Out[95]=
```

Factorial	$\{\cdots_{10}\}$
Random number	$\{\cdots_{10}\}$
Natural Logarithm	$\{\cdots_{10}\}$

Figure 3-35. *Dataset values by Mathematica due to large contents*

If we click on the rows, we should get the values for the corresponding row.

Functions by Column or Row

Another approach is to directly apply a function to our values of a column and we can specify a rule of transformation. For example, we can round to the smallest integer greater than or equal to all the values on the column Natural Logarithm. Figure 3-36 shows the output.

```
In[96]:= DataNumbr[All,{"Natural Logarithm" → Ceiling}](*The same can be
done using the index number of the columns, DataNumbr *)
Out[96]=
```

Factorial	Random number	Natural Logarithm
1	0.983625	0
2	0.252122	1
6	0.115411	2
24	0.652804	2
120	0.355895	2
720	0.230432	2
5040	0.945289	2
40320	0.830881	3
362880	0.197961	3
3628800	0.202727	3

Figure 3-36. *Ceiling function applied as a rule*

We can apply the square root to the first row. Map can also be used to apply functions to rows. Figure 3-37 shows the output generated.

```
In[97]:= DataNumbr[1,Sqrt] (* Map[Sqrt,DataNumbr[1;;2,All]] can also do the
work for the first 2 rows*)
Out[97]=
```

Factorial	1
Random number	0.991779
Natural Logarithm	0

Figure 3-37. *Output generated from the earlier code*

On the occasion that we want to apply a function to a defined level, we can use MapAt. MapAt has the form MapAt[f, "expr", {i, j, ...}], where {i, j} means the level of the position, as shown in Figure 3-38.

```
In[98]:=MapAt[Exp,DataNumbr,{1}](*for first position of row 1 only*)
(*Double semi-colon can be used to define from row to row, try using 4 ;;
6. Caution you might get big numbers *)
Out[98]=
```

Factorial	Random number	Natural Logarithm
2.71828	2.67413	1
2	0.252122	0.693147
6	0.115411	1.09861
24	0.652804	1.38629
120	0.355895	1.60944
720	0.230432	1.79176
5040	0.945289	1.94591
40320	0.830881	2.07944
362880	0.197961	2.19722
3628800	0.202727	2.30259

Figure 3-38. *Exponentiation for the first row only with MapAt*

Occasionally we might encounter duplicate data, and this can make it hard to understand our data, especially if something goes wrong. One approach can be to remove an entire row or column, as we saw in previous sections; but as an alternative, we can use built-in functions that can do the job. The function DeleteDuplicates is the most common. DeleteCases can be used too, but it removes data that match a pattern, in contrast to DeleteDuplicates. Let us create a dataset for our example.

```
In[99]:= Sales=Dataset@
{
<|"Id" → 1, "Product" → "PC", "Price" → "800 €", "Sale Month" →
"January"|>,
<|"Id" → 2, "Product" → "Smart phone", "Price" → "255 €", "Sale Month"
→ "January"|>,
<|"Id" → 3, "Product" → "Anti-Virus", "Price" → "100 €", "Sale Month" →
"March"|>,
<|"Id" → 4, "Product" → "Earphones", "Price" → "78 €", "Sale Month" →
"February"|>,
<|"Id" → 5, "Product" → "PC", "Price" → "809 €", "Sale Month" →
"March"|>,
<|"Id" → 5, "Product" → "PC", "Price" → "809 €", "Sale Month" →
"March"|>,
<|"Id" → 6, "Product" → "Radio", "Price" → "60 €", "Sale Month" →
"January"|>,
<|"Id" → 7, "Product" → "PC", "Price" → "700 €", "Sale Month" →
"February"|>,
<|"Id" → 8, "Product" → "Mouse", "Price" → "100 €", "Sale Month" →
"March"|>,
<|"Id" → 9, "Product" → "Keyborad", "Price" → "125 €", "Sale Month" →
"January"|>,
<|"Id" → 10, "Product" → "USB 64gb", "Price" → "90 €", "Sale Month" →
"March"|>,
<|"Id" → 11, "Product" → "LED Screen", "Price" → "900 €", "Sale Month"
→ "February"|>,
<|"Id" → 11, "Product" → "LED Screen", "Price" → "900 €", "Sale Month"
→ "February"|→>
}
Out[99]=
```

Id	Product	Price	Sale Month
1	PC	800 €	January
2	Smart phone	255 €	January
3	Anti-Virus	100 €	March
4	Earphones	78 €	February
5	PC	809 €	March
5	PC	809 €	March
6	Radio	60 €	January
7	PC	700 €	February
8	Mouse	100 €	March
9	Keyborad	125 €	January
10	USB 64gb	90 €	March
11	LED Screen	900 €	February
11	LED Screen	900 €	February

Figure 3-39. *Dataset example for duplicate data*

As can be seen in Figure 3-39, in the dataset we have two rows that are duplicated, rows with ID numbers 5 and 11. The function DuplicateFreeQ can detect whether the dataset appears to have duplicates. The function returns False when we have duplicate data and True when we do not. It can be applied straight to the dataset or we can detect the rows that appear to be duplicated.

Let us check if we have duplicates from rows 1 through 7.

```
In[100]:= DuplicateFreeQ[Sales[1;;7,All]]
Out[100]= False
```

We have found programmatically that we have duplicate data in the dataset. Another form is to check if we have duplicates by column.

```
In[101]:= Sales[All,{"Id"}]@DuplicateFreeQ
Out[101]= False
```

Now that we have found that indeed we have duplicates, to delete duplicates the function DeletDuplicates is used. It can be applied to the dataset or to a particular column or row as a function. The output generated is shown in Figure 3-40.

```
In[102]:= DeleteDuplicates[Sales] (*Datas[All,{"ID"}]@DuplicateFreeQ*)
Out[102]=
```

Id	Product	Price	Sale Month
1	PC	800 €	January
2	Smart phone	255 €	January
3	Anti-Virus	100 €	March
4	Earphones	78 €	February
5	PC	809 €	March
6	Radio	60 €	January
7	PC	700 €	February
8	Mouse	100 €	March
9	Keyborad	125 €	January
10	USB 64gb	90 €	March
11	LED Screen	900 €	February

Figure 3-40. *Dataset without duplicates*

An alternative is to use GroupBy to identify which data is duplicated in our dataset.

```
In[103]:= GroupBy[Sales,"Id"]
Out[103]=
```

Notice in Figure 3-41, the repeated data are stacked together.

	Id	Product	Price	Sale Month
1	1	PC	800 €	January
2	2	Smart phone	255 €	January
3	3	Anti–Virus	100 €	March
4	4	Earphones	78 €	February
5	5	PC	809 €	March
	2 total ›			
6	6	Radio	60 €	January
7	7	PC	700 €	February
8	8	Mouse	100 €	March
9	9	Keyborad	125 €	January
10	10	USB 64gb	90 €	March
11	11	LED Screen	900 €	February
	2 total ›			

Figure 3-41. *Dataset grouped by duplicates*

Customizing a Dataset

Datasets can be customized depending on how we want to show our data, working with datasets can be personalized by preferences. To explore this in the next block, we will load example data from the Wolfram reference servers to discover the ways you can personalize data for your needs. When loading data from the server, depending on your internet connection, it might pop up a loading frame trying to access the Wolfram servers.

Let us load the data by using ExampleData and then choosing statistics of animal weights and converting the list into a dataset. By using the option MaxItem, we can display how many rows or columns to exhibit of the dataset. We choose to show the first four rows and the first three columns. When showing the dataset, scroll bars will appear on the left side and topside. Use them to move over the dataset. Alternatively, we can choose to align the contents on the left, center, or right side. In Figure 3-42, only the left scrollbars appear.

```
In[104]:= AnimalData= ExampleData[{"Statistics","AnimalWeights"}];
Dataset[ AnimalData, MaxItems → {4,3}, Alignment → Center] (*To align a
sole column, Alignment → {"Col_name" → Left} *)
Out[104]=
```

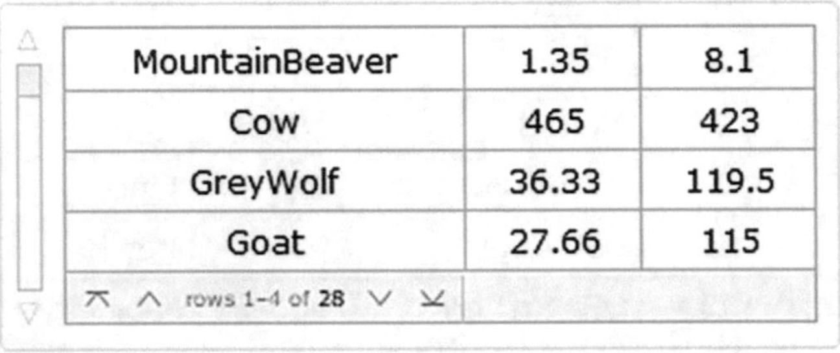

MountainBeaver	1.35	8.1
Cow	465	423
GreyWolf	36.33	119.5
Goat	27.66	115

rows 1–4 of 28

Figure 3-42. *Animal dataset*

To color the contents of the dataset, the option Background is used and the colors the notation {row, col} is preserved. To paint the whole data, enter only the color. To paint by row or column, enter the colors as a nested list—that is, {{"color_row1", "color_row2", ... }, {"color_col1", "color_col2", ... }}. Mixing colors can also be done by nesting the nested color. For specific values, the position of the values would need to be entered. In the next example, we will color the first two columns only like in Figure 3-43.

For particular values, the position of the values would need to be entered. Another option is the size of the items, and this is controlled with ItemSize option. If we want to edit the same options but with headers, we would use HeaderAlignment for placing text to left, center, or right; HeaderSize for the size of the titles; and HeaderStyle for the style of the font.

```
In[105]:= Dataset[AnimalData,MaxItems → {4,3}, Background → {{None},
{LightBlue,LightYellow}},ItemSize → {12}]
Out[105]=
```

MountainBeaver	1.35	8.1
Cow	465	423
GreyWolf	36.33	119.5
Goat	27.66	115

⋀ ⋀ rows 1–4 of 28 ⋁ ⋁

Figure 3-43. *Columns 1 and 2 colored*

For particular values, the position of the values would need to be entered. Another option is the size of the items, and this is controlled with ItemSize option. If we want to edit the same options but with headers, we would use HeaderAlignment for placing the text left, center, or right; HeaderSize for the size of the titles; and ItemStyle for the style of the font of the items. Figure 3-44 shows the data set in bold style.

```
In[106]:= Dataset[AnimalData,MaxItems → {4,3}, Background → {{4,3} →
Yellow}, ItemSize → {12}, ItemStyle → Bold]
Out[106]=
```

MountainBeaver	**1.35**	**8.1**
Cow	**465**	**423**
GreyWolf	**36.33**	**119.5**
Goat	**27.66**	**115**

⋀ ⋀ rows 1–4 of 28 ⋁ ⋁

Figure 3-44. *Dataset with bold style*

Another useful option is HiddenItems, which hides items that do not want to be shown. Therefore, to hide row 1 and column 1 would be. Columns can be hidden with their associated label, like: HiddenItems → {"row #", "col #"}. Figure 3-45 shows the form of suppressed rows and columns in the dataset. For specific values, then nest the position of the value try HiddenItems → {{2,3}}.

```
In[107]:= Dataset[AnimalData,MaxItems → {4,3}, HiddenItems → {1,1}]
Out[107]=
```

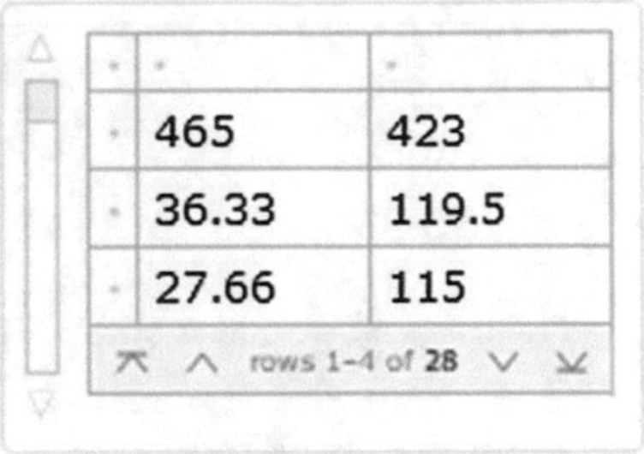

465	423
36.33	119.5
27.66	115

rows 1–4 of 28

Figure 3-45. *Column 1 and row 1 suppressed*

We can add headers to each of the columns in our new data set with the Query command. To rename the columns, the same procedure is applied; the new names would be ruled to the old names—that is, "New name" → "Animal Name," as shown in Figure 3-46.

```
In[108]:= Query[All,<|"Animal Name" → 1, "Body Weight" → 2, "Brain
Weight" → 3|>]@Dataset[AnimalData]
(* for display motives we put row 7 to 9, use All for the whole data set *)
(* or "symbol_of_the_dataset" [All,<|"Animal Name" → 1, "Body Weight" →
2, "Brain Weight" → 3|>] *)
Out[108]=
```

Animal Name	Body Weight	Brain Weight
MountainBeaver	1.35	8.1
Cow	465	423
GreyWolf	36.33	119.5
Goat	27.66	115
GuineaPig	1.04	5.5
Diplodocus	11700	50
AsianElephant	2547	4603
Donkey	187.1	419
Horse	521	655
PotarMonkey	10	115
Cat	3.3	25.6
Giraffe	529	680
Gorilla	207	406
Human	62	1320
AfricanElephant	6654	5712
Triceratops	9400	70
RhesusMonkey	6.8	179
Kangaroo	35	56
GoldenHamster	0.12	1
Mouse	0.023	0.4

⊼ ∧ rows 1–20 of 28 ∨ ⊻

Figure 3-46. *Animal dataset with added column headers*

Generalization of Hash Tables

A hash table is an associative data structure, which allows the storage of data and, in turn, the rapid retrieval of elements (values) from objects called keys. Hash tables can be implemented inside arrays, where the main components are the key and the value. The way to search for an element in the array is by using a hash function, which maps the keys to the pairs of values, which gives us the place where it is in the array (index).

125

In other words, the hash function searches for a certain key, evaluates that key, and returns an index. This process is known as hashing. Figure 3-47 shows a representative schema of a hash table.

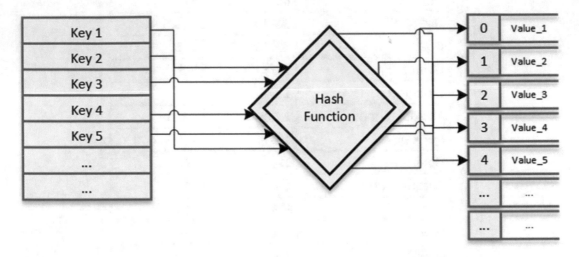

Figure 3-47. *Graphic representation of hash table*

Inside the hash table, the number of keys and values can go on and on. This is one of the reasons hash tables are very useful, because it can store large amounts of information. Inside the Wolfram Language, associations can represent hash tables. This is primarily because associations are an abstract data structure with its fundamentals components are keys and values, just like a hash table. This combines the structure of an associative array and an indexed list, more like a nest of hash arrays. With a crucial property that associations are immutable, this means that each association-type object is unique, and that the reference of one association has no link to another, even though they are referenced to the same symbol.

Just like we saw common commands with associations, more special commands are available. Let's first create an association. Nested associations are defined as associations that have associations within—in other words, a key that points to a bucket of values that correspond to keys that have other values inside (Figure 3-48).

```
In[109]:=
Asc=<|"User"->
      <|"Edgar"->
            <|"id"-> 01,
                "Parameters"->
                    <|"Active"-> True,"Region"-> "LA","Internet Traffic"->
"1 GB"|>|>,
        <|"Anya"->
              <|"id"-> 02,
                  "Parameters"->
                    <|"Active"-> False,"Region"-> "MX","Internet Traffic"->
"3 GB"|>|>|>|>|>;
Dataset[%]
Out[109]=
```

		id	Parameters		
			Active	Region	Internet Traffic
User	Edgar	1	True	LA	1 GB
	Anya	2	False	MX	3 GB

Figure 3-48. *Nested associations in the dataset format*

Executing operations like accessing items, updating values, and deleting is supported with the commands associated to keys and values. Remember that Keys returns the keys of the association an Values the values. Inside a nested association, Keys only work on the surface level. This is seen in the next code.

```
In[110]:= Keys[Asc]
Out[110]= {User}
```

Applying Keys only returns the key user. To see the keys inside a nested association, the command Keys needs to be applied to more deep levels. This is achieved with Map and specifying the sublevel only.

```
In[111]:= Map[Keys,Asc,#]&/@{{0},{1},{2}}//Column
Out[111]= {User}
<|User->{Edgar,Anya}|>
<|User-><|Edgar->{id,Parameters},Anya->{id,Parameters}|>|>
```

As seen on the surface level (0), the key is User. The next sublevel has the keys Edgar and Anya, and the last level has the keys ID and parameters for each of the keys Edgar and Anya. With MapIndexed we can look inside the whole association and apply Keys into sublevels to show the predecessor of the keys.

```
In[112]:=
Print["Level 0: "<>ToString@MapIndexed[Keys,Asc,{0}]]
Print["Level 1: "<>ToString@MapIndexed[Keys,Asc,{1}]]
Print["Level 2: "<>ToString@MapIndexed[Keys,Asc,{2}]]
Level 0: {{}[User]}
Level 1: <|User -> {{Key[User]}[Edgar], {Key[User]}[Anya]}|>
Level 2: <|User -> <|Edgar -> {{Key[User], Key[Edgar]}[id], {Key[User],
Key[Edgar]}[Parameters]}, Anya -> {{Key[User], Key[Anya]}[id], {Key[User],
Key[Anya]}[Parameters]}|>|>
```

At level 0, only the key User exists, and the predecessor is {}. At level 1 the predecessor User and the keys Edgar and Anya are values of the key User, and at level 2 the predecessor keys are Edgar/Anya and User for the keys ID and Parameters. In other words, the expression {Key[User], Key[Anya]}[id], means that ID corresponds to the key Anya and Anya to the key User, and so on. This is also useful because it means that access to a value or values of a key is done with the operator form applied to the association specifying the keys.

```
In[113]:=
Asc["User"]["Edgar"]["id"](*{Key[User],Key[Anya]}[id],*)
Out[113]=0
```

As shown, we get the value that corresponds to the key ID inside Edgar inside User. To see a graphical representation of the previous expression, we can use MapIndexed to label the positions of the keys and dataset applied, for example, in sublevel 4 (Figure 3-49)

```
In[114]:= Dataset@MapIndexed[Framed[Labeled[#2,#1],FrameMargins-
>0,RoundingRadius->5]&,Asc,{4}](*Try changin the number to see how the
expression changes*)
Out[114]=
```

User		id		Parameters
User	Edgar	1	Active	{Key["User"], Key["Edgar"], Key["Parameters"], Key["Active"]} True
			Region	{Key["User"], Key["Edgar"], Key["Parameters"], Key["Region"]} "LA"
			Internet Traffic	{Key["User"], Key["Edgar"], Key["Parameters"], Key["Internet Traffic"]} "1 GB"
	Anya	2	Active	{Key["User"], Key["Anya"], Key["Parameters"], Key["Active"]} False
			Region	{Key["User"], Key["Anya"], Key["Parameters"], Key["Region"]} "MX"
			Internet Traffic	{Key["User"], Key["Anya"], Key["Parameters"], Key["Internet Traffic"]} "3 GB"

Figure 3-49. *Dataset representation marking the keys inside the nested association*

Each box contains the values of the predecessor key. This is why 1 GB corresponds to {Key[User],Key[Edgar],Key[Parameters],Key[Internet Traffic]}. To see the whole expression, the level of specification is Infinity (Figure 3-50).

```
In[115]:=
MapIndexed[Framed[Labeled[#2,#1,ImageMargins->0,Spacings->0],FrameMargins->0,
RoundingRadius->5]&,Asc,Infinity]
Out[115]=
```

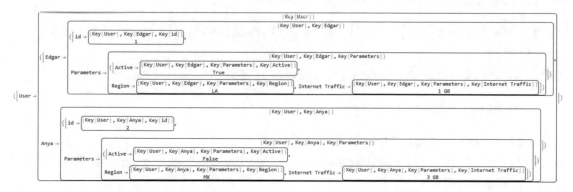

Figure 3-50. *Framed levels of the keys in a nested association*

Values uses the same approach as with Keys. To test if key exists, use KeyExistQ; this returns true if the key exists. Otherwise it is false. To test inside deeper levels, use Map.

```
In[116]:= {KeyExistsQ[Asc,"User"],Map[KeyExistsQ["Anya"],Asc,{1}],
Map[KeyExistsQ["Anya"],Asc,{2}]}
Out[116]= {True,<|User->True|>,<|User-><|Edgar->False,Anya->False|>|>}
```

Another way to test whether key in a particular form exists inside an association, use KeyMemberQ—for example, if there is a string pattern key.

```
In[117]:= KeyMemberQ[Asc["User"]["Anya"],_String]
Out[117]= True
```

To test if a value exists given a key, use Lookup.

```
In[118]:= Lookup[Asc["User"]["Anya"],"Parameters"]
Out[118]= <|Active->False,Region->MX,Internet Traffic->3 GB|>
```

To select a key based on criteria, use KeySelect.

```
In[119]:= KeySelect[Asc["User"]["Anya"],StringQ]
Out[119]= <|id->2,Parameters-><|Active->False,Region->MX,Internet
Traffic->3 GB|>|>
```

Or simply use KeyTake to grab a particular key.

```
In[120]:= KeyTake[Asc["User"]["Anya"]["Parameters"],{"Region","Internet
Traffic"}]
Out[120]= <|Region->MX,Internet Traffic->3 GB|>
```

To remove a key, use KeyDrop.

```
In[121]:= KeyDrop[Asc["User"],"Edgar"]
Out[121]= <|Anya-><|id->2,Parameters-><|Active->False,Region->MX,Internet
Traffic->3 GB|>|>|>
```

To assign a new value, the value associated with the key is assigned with the new value.

```
In[122]:= Asc["User"]["Edgar"]["Parameters"]["Region"]="CZ"
Out[122]= CZ
```

Passing this into a dataset, we can look for the new assigned value (Figure 3-51).

```
In[123]:= Dataset[Asc]
Out[123]=
```

		id	Parameters		
			Active	Region	Internet Traffic
User	Edgar	1	True	CZ	1 GB
	Anya	2	False	MX	3 GB

Figure 3-51. *Dataset with the region value changed to CZ*

To add a key and a value to the association, we can Insert the new expression by specifying the position to insert it with the key (Figure 3-52).

```
In[124]:= Insert[Asc["User"],"Alexandra"-><|"id"->0,"Parameters"->
<|"Active"->False,"Region"->"RS","Internet Traffic"->"12 GB"|>|>,
Key["Edgar"]]//Dataset
Out[124]=
```

	id	Parameters		
		Active	Region	Internet Traffic
Alexandra	0	False	RS	12 GB
Edgar	1	True	CZ	1 GB
Anya	2	False	MX	3 GB

Figure 3-52. *New row added by the key position*

CHAPTER 4

Import and Export

In this chapter, we will review the import and export of data. Here we will see the Wolfram Language commands to import and export data. We will review what type of formats Mathematica supports for both import and export.

Experimental data can come from different sources. The way to process this external data is to import it through Wolfram Language. Data that has been calculated or obtained externally can be transferred to Mathematica as well as exported for use on other platforms. However, Mathematica has tools to handle different types of data (numbers, text, audio, graphics, and images). We will focus on working with numerical and categorical data, which are probably the most frequent type of data used for analysis.

Importing data into Mathematica from multiple sources allows loading data into a notebook for analysis. Numerous import formats are supported by the Wolfram Language; to see which formats are supported, type the dollar symbol ($) accompanied with the command ImportFormats. Currently Mathematica supports 226 file formats.

```
In[1]:= Short[$ImportFormats,5]
```

```
Out[1]//Short= {3DS,ACO,Affymetrix,AgilentMicroarray,AIFF,ApacheLog,ArcGRID
,AU,AVI,Base64,BDF,Binary,Bit,BMP,BSON,Byte,BYU,BZIP2,CDED,<<189>>,VTK,WARC
,WAV,Wave64,WDX,WebP,WLNet,WMLF,WXF,XBM,XHTML,XHTMLMathML,XLS,XLSX,XML,XPOR
T,XYZ,ZIP}
```

And we can see in the list a lot of formats: audio, image, text, etc. But we will focus on the text-based formats and others. To import any kind of format file, the command Import is used. Import receives two arguments, which is the path of the file and options. Options can vary between file format, elements, and other types of objects in Mathematica, like cloud, local, etc. To select a file path, head to the toolbar, then go to Insert ▶File Path… A file explorer should appear, and then search the file you would like to import and select it. The path will be enclosed in apostrophes like a string.

© Jalil Villalobos Alva 2021
J. Villalobos Alva, *Beginning Mathematica and Wolfram for Data Science*,
https://doi.org/10.1007/978-1-4842-6594-9_4

As you will notice, there is also another option named File in the Insert menu. In contrast to File Path, File will introduce the contents of the file directly without receiving prior formatting from Mathematica. File is better suited for importing notebooks or other Wolfram formats.

Let us see how to transfer a simple text file.

```
In[2]:= Import["C:\\Users\\My-pc\\Desktop\\HelloWorld.txt"]
Out[2]= Hello World!
```

We have imported our first file. Mathematica will recognize it based on the file extension and then import it automatically. If we import a file with no file extension, but we know the type of format used in the file, we can choose the right format as an option.

```
In[3]:= Import["C:\\Users\\My-pc\\Desktop\\HelloWorld.txt","Text"]
Out[3]= Hello World!
```

Importing Files

CSV and TSV Files

In this section we will focus on how to import files into Mathematica. In the examples we will work with comma-separated value (CSV) files, tab-separated value (TSV) files, and Excel spreadsheet style files. CSV and TSV files are files that include text and numeric values. In CSV files, fields are separated by a comma; each row is one line record. Whereas in TSV files, each record is separated with a tab space.

With Import, we can import TSV or CSV files with the .tsv or .csv file extension, respectively. We will first import a normal CSV file by introducing the file path and then the CSV option.

```
In[4]:=Import["C:\\Users\\My-pc\\Desktop\\Grocery_List.csv","CSV"]
Out[4]= {{id,grocery item,price,sold items,sales per day},{1,milk,4$,4,4
Jun 2019},{2,butter,3$,2,6 Jun 2019},{3,garlic,2$,1,7 Jun
2019},{4,apple,2$,4,1 Jun 2019},{5,orange,3$,5,8 Jun 2019},{6,orange
juice,5$,2,8 Jun 2019},{7,cheese,5$,2,6 Jun 2019},{8,cookies,2$,5,9 Jun
2019},{9,grapes,4$,3,21 Jun 2019},{10,potatoe,2$,5,26 Jun 2019}}
```

Now that the contents of the file are imported, depending on the format of the contents, the data is presented as a nested list or not. Rows are represented as elements of the nested list and columns as the elements of the whole list.

When importing data, parts of the data can be imported—that is, if we only need a row or a column.

```
In[5]:=Import["C:\\Users\\My-pc\\Desktop\\Grocery_List.csv",{"Data",5;;10}]
Out[5]= {{4,apple,2$,4,1 Jun 2019},{5,orange,3$,5,8 Jun 2019},{6,orange
juice,5$,2,8 Jun 2019},{7,cheese,5$,2,6 Jun 2019},{8,cookies,2$,5,9 Jun
2019},{9,grapes,4$,3,21 Jun 2019}}
```

In the previous example we imported data from row 5 to row 10.

When we are only interested in single values only, then we can use the following form.

```
In[6]:= Import["C:\\Users\\My-pc\\Desktop\\Grocery_List.csv",{"Data",6,2}]
Out[6]= orange
```

Depending of the maximum bytes of the expression, Mathematica will truncate the imported data and show you a suggestion box of a simplified version of the whole data. To see the maximum byte size, go to, Edit ➤ Advanced tab, and in "Maximum output size before truncation," enter the new number of bytes before truncation. This preference applies to every output expression in Mathematica.

To import TSV files, we use the same approach. With the short command, we can show a part of the data, just in case the data is very large.

```
In[7]:= Short[Import[
"C:\\Users\\My-pc\\Desktop \\Color_table.tsv","TSV"]] (*Rest, to view the
remain *)
Out[7]//Short= {{number,color},{1,red},<<7>>,{9,magenta},{10,brown}}
```

Consequently, we see that in the result, a seven appears among the elements of the imported file. This is because it contains seven elements that are not shown. Now that we know how to import CSV and TSV files, imported data can be displayed in table format with Grid or TableForm.

```
In[8]:= Import[
"C:\\Users\\My-pc\\Desktop\\Grocery_List.csv","CSV"];
Grid[%]
```

```
Out[8]=
```

id	grocery	item	price	sold items	sales per	day
1	milk		4$	4	4	Jun 2019
2	butter		3$	2	6	Jun 2019
3	garlic		2$	1	7	Jun 2019
4	apple		2$	4	1	Jun 2019
5	orange		3$	5	8	Jun 2019
6	orange juice		5$	2	8	Jun 2019
7	cheese		5$	2	6	Jun 2019
8	cookies		2$	5	9	Jun 2019
9	grapes		4$	3	21	Jun 2019
10	potatoe		2$	5	2	Jun 2019

And now that we have assigned a name to the imported data, the contents can now be treated as a list. Parts of our data can be extracted, as we will see in later chapters.

XLSX Files

In the next example we will expose how to import data and display data as a spreadsheet and how to transform it into a dataset. For exemplification purposes, we will use the XLSX grocery list file rather than the CSV file. To start, we need to first import our data.

```
In[9]:=
path="C:\\Users\\My-pc\\Desktop \\Grocery_List.xlsx";
Import[path,"Data"]
Out[9]=
{{{id,grocery item, price, sold items,sales per day},{1.,milk,4
$,4., 4-Jun-2019},{2.,butter,3$,2., 6-Jun-2019},{3.,garlic,2 $,1.,
7-Jun-2019},{4.,apple,2 $,4., 1-Jun-2019},{5.,orange,3 $,5., 8-Jun-
2019},{6.,orange juice,5 $,2., 8-Jun-2019},{7.,cheese,5 $,2., 6-Jun-
2019},{8.,cookies,2 $,5., 9-Jun-2019},{9.,grapes,4 $,3., 21-Jun-
2019},{10.,potatoe,2 $,5., 26-Jun-2019}}}
```

As can be seen, the data imported appear as a nested list; that is because Excel files can have multiple sheets inside a file. For this case we have only one sheet. To see the number of sheets and name of the sheets, use SheetCount and Sheets, respectively.

```
In[10]:= Import[path,#]&/@{"SheetCount","Sheets"}
Out[10]={1,{Grocery_List}}
```

To show data as a spreadsheet, we use the command TableView (Figure 4-1). To select a sheet the next format is used as an option: {"Data", # of sheet}. And to select a character encoding, use the CharacterEncoding option. Also costume rows or columns can be imported preserving the format: {"Data", # of sheet, # row, # column}.

```
In[11]:= TableView[Import[path,{"Data",1},CharacterEncoding→"UTF-8"]]
Out[11]=
```

	1	2	3	4	5
1	id	grocery item	price	sold items	sales per day
2	1.	milk	4 $	4.	4-Jun-2019
3	2.	butter	3$	2.	6-Jun-2019
4	3.	garlic	2 $	1.	7-Jun-2019
5	4.	apple	2 $	4.	1-Jun-2019
6	5.	orange	3 $	5.	8-Jun-2019
7	6.	orange juice	5 $	2.	8-Jun-2019
8	7.	cheese	5 $	2.	6-Jun-2019
9	8.	cookies	2 $	5.	9-Jun-2019
10	9.	grapes	4 $	3.	21-Jun-2019
11	10.	potatoe	2 $	5.	26-Jun-2019
12					
13					
14					

Figure 4-1. *Spreadsheet view with TableView command*

Note   With "Data," import the data as a nested list.

As you may notice, we can now see our data in spreadsheet format. Now with TableView we can view our data like in spreadsheet software, with selection tools, scrollbars, and text editing of the contents. However, one of the downsides is that with TableView we cannot directly access the contents of the file; also neither calculation can be performed. To do the latter, we can transform it into a dataset.

We can convert data into a dataset for better handling in Mathematica. By typing the option "Dataset" instead of "Data," the imported file becomes a dataset but without headers (as shown in Figure 4-2). To add the headers, use the HeaderLines option, and to choose the specification of header by row or column type HeadLines → { # row, # column }. The file used is Grocery List 2.xlxs.

```
In[12]:= file=
"C:\\Users\\My-pc\\Desktop\\Grocery_List_2.xlsx";
Import[file,{"Dataset",1},HeaderLines→1]
Out[12]=
```

id	grocery item	price	sold items
1.0	milk	4$	4.0
2.0	butter	3$	
3.0	garlic	2$	1.0
4.0	apple	2$	4.0
5.0	orange		5.0
6.0	orange juice	5$	2.0
7.0	cheese	5$	2.0
8.0	cookies		5.0
9.0	grapes	4$	
10.0	potatoe	2$	5.0

Figure 4-2. *Incomplete Grocery List dataset*

As you may notice, we import incomplete data. To treat empty spaces, the EmptyFiled is implemented as a rule of transformation. If the data has empty spaces and no rule is expressed, the spaces will be treated as empty strings. Figure 4-3 shows the output.

```
In[13]:= Import[file,{"Dataset",1},"EmptyField"→ "NaN",HeaderLines→1]
Out[13]=
```

id	grocery item	price	sold items
1.0	milk	4$	4.0
2.0	butter	3$	NaN
3.0	garlic	2$	1.0
4.0	apple	2$	4.0
5.0	orange	NaN	5.0
6.0	orange juice	5$	2.0
7.0	cheese	5$	2.0
8.0	cookies	NaN	5.0
9.0	grapes	4$	NaN
10.0	potatoe	2$	5.0

Figure 4-3. *NaN-filled dataset*

JSON Files

The JavaScript Object Notation (JSON) file extension is a data representation file. JSON files are used to store data as an ordered list of values, and each list is constituted by a collection of value pairs. To import a JSON file, we can specify the two options "JSON" or "RawJSON."

```
In[14]:= Json=Import[
"C:\\Users\\My-pc\\Desktop\\Sports_cars.json","JSON"]
Out[14]=
{{Model→Enzo Ferrari,Year→2002,Cylinders→12,Horsepower HP→660,Weight
Kg→1255},{Model→Koenigsegg CCX,Year→2000,Cylinders→8,Horsepower
HP→806,Weight Kg→1180},{Model→Pagani Zonda,Year→2002,Cylinders→12,Horse
power HP→558,Weight Kg→1250},{Model→McLaren Senna,Year→2019,Cylinders→8
,Horsepower HP→800,Weight Kg→1309},{Model→McLaren 675 LT,Year→2015,Cyli
nders→8,Horsepower HP→675,Weight Kg→1230},{Model→Bugatti Veyron,Year→20
06,Cylinders→16,Horsepower HP→1001,Weight Kg→1881},{Model→Audi R8 Spyde
```

r,Year→2010,Cylinders→10,Horsepower HP→525,Weight Kg→1795},{Model→Aston Martin Vantage,Year→2009,Cylinders→8,Horsepower HP→926,Weight Kg→1705},{Model→Maserati Gran Turismo,Year→2010,Cylinders→8,Horsepower HP→405,Weight Kg→1955},{Model→Lamborghini Aventador S,Year→2017,Cylinders→12,Horsepower HP→740,Weight Kg→1740}}

Given the nature of the structure of the JSON file, when importing them, Mathematica recognizes each structure and interprets each key to its values. As we see in the previous output, keys correspond to Model, Year, Cylinders, Horsepower, and Weight, and each key has its values. Everything said so far explains that all records are contained in a nested list. This leads us to the conclusion that if we want to present it in the form of a dataset, we could not directly apply Association, and Association suppresses repeated keys. We will have to create an association for each record since it is a nested list, and this we will achieve with Map, specifying the depth level of the Association command. This is shown here.

```
In[15]:= Map[Association,Json,1]
Out[15]= {<|Model→Enzo Ferrari,Year→2002,Cylinders→12,Horsepower
HP→660,Weight Kg→1255|>,<|Model→Koenigsegg CCX,Year→2000,Cylinders→8,
Horsepower HP→806,Weight Kg→1180|>,<|Model→Pagani Zonda,Year→2002,Cyli
nders→12,Horsepower HP→558,Weight Kg→1250|>,<|Model→McLaren Senna,Yea
r→2019,Cylinders→8,Horsepower HP→800,Weight Kg→1309|>,<|Model→McLaren
675 LT,Year→2015,Cylinders→8,Horsepower HP→675,Weight
Kg→1230|>,<|Model→Bugatti Veyron,Year→2006,Cylinders→16,Horsepower
HP→1001,Weight Kg→1881|>,<|Model→Audi R8 Spyder,Year→2010,Cylinders→1
0,Horsepower HP→525,Weight Kg→1795|>,<|Model→Aston Martin Vantage,Year
→2009,Cylinders→8,Horsepower HP→926,Weight Kg→1705|>,<|Model→Maserati
Gran Turismo,Year→2010,Cylinders→8,Horsepower HP→405,Weight
Kg→1955|>,<|Model→Lamborghini Aventador S,Year→2017,Cylinders→12,Horsep
ower HP→740,Weight Kg→1740|>}
```

As we can see we already have each record as an association, and now we can convert it to a dataset, as can be seen in Figure 4-4.

```
In[16]:= Dataset[%]
Out[16]:=
```

Model	Year	Cylinders	Horsepower HP	Weight Kg
Enzo Ferrari	2002	12	660	1255
Koenigsegg CCX	2000	8	806	1180
Pagani Zonda	2002	12	558	1250
McLaren Senna	2019	8	800	1309
McLaren 675 LT	2015	8	675	1230
Bugatti Veyron	2006	16	1001	1881
Audi R8 Spyder	2010	10	525	1795
Aston Martin Vantage	2009	8	926	1705
Maserati Gran Turismo	2010	8	405	1955
Lamborghini Aventador S	2017	12	740	1740

Figure 4-4. *Cars dataset*

We can now handle a JSON file as a dataset. However, there is another way to do it without the need for so much calculation. This is achieved when importing the file. We must import it as RawJson, because with RawJson the Wolfram Language identifies and imports each record as a list of associations rather than a sole nested list, as shown here. This is because of the nature of key and value of the JSON file extension.

```
In[17]:= Import["C:\\Users\\My-pc\\Desktop\\Sports_cars.json","RawJSON"]
Out[17]= {<|Model→Enzo Ferrari,Year→2002,Cylinders→12,Horsepower
HP→660,Weight Kg→1255|>,<|Model→Koenigsegg CCX,Year→2000,Cylinders→8,
Horsepower HP→806,Weight Kg→1180|>,<|Model→Pagani Zonda,Year→2002,Cyli
nders→12,Horsepower HP→558,Weight Kg→1250|>,<|Model→McLaren Senna,Yea
r→2019,Cylinders→8,Horsepower HP→800,Weight Kg→1309|>,<|Model→McLaren
675 LT,Year→2015,Cylinders→8,Horsepower HP→675,Weight
Kg→1230|>,<|Model→Bugatti Veyron,Year→2006,Cylinders→16,Horsepower
HP→1001,Weight Kg→1881|>,<|Model→Audi R8 Spyder,Year→2010,Cylinders→1
0,Horsepower HP→525,Weight Kg→1795|>,<|Model→Aston Martin Vantage,Year
→2009,Cylinders→8,Horsepower HP→926,Weight Kg→1705|>,<|Model→Maserati
Gran Turismo,Year→2010,Cylinders→8,Horsepower HP→405,Weight
Kg→1955|>,<|Model→Lamborghini Aventador S,Year→2017,Cylinders→12,Horsep
ower HP→740,Weight Kg→1740|>}
```

As we see now, the file is imported as an association in each record and we can proceed to convert it into a dataset.

```
In[18]:=Cars=Dataset[%];
```

As a complement, once the data is imported, we can perform operations on the dataset, such as ordering the models by year from low to high.

```
In[19]:=Cars[SortBy[#Year& ]];
```

Note   The previous example is also possible using the query command. (Query [SortBy[#Year &]][Cars]).

Web Data

On the other hand, web data is also supported with Import. Instead of inserting the file path, the URL site is inserted as the argument of the Import command. In the next example we will import a simple text file from the National Oceanic and Atmospheric Administration (NOAA). The text file will contain the list of country codes use for the Integrated Global Radiosonde Archive (IGRA). The parent directory where files are located can be found at https://www1.ncdc.noaa.gov/pub/data/igra/, but we will only import the country list file. You need internet connection to make this work.

```
In[20]:= Short[Import["https://www1.ncdc.noaa.gov/pub/data/igra/igra2-
country-list.txt","HTML"]]
Out[20]//Short=
 AC Antigua and Barbuda AE United Ara ... Yemen ZA Zambia ZI Zimbabwe ZZ
Ocean
```

As you can see the file is a plain text file, but we can change how the data is imported by inserting a file format as an option. We can import it as a CSV file, for instance.

```
In[21]:= Short[Import["https://www1.ncdc.noaa.gov/pub/data/igra/igra2-
country-list.txt","CSV"]]
Out[21]//Short=
{{AC Antigua and Barbuda},{AE United Arab Emirates},<<216>>,{ZZ Ocean}}
```

This is useful when we try to make computation with the data imported. As an alternative we can use URL commands to check the status of an online file, and then download it.

To check the status of the online file, use URLRead. In the occasion the file is online, you should get an http response object like the one in the Figure 4-5. This approach can even be done before importing data; with this you can be sure that the content is online.

```
In[22]:= URLRead["https://www1.ncdc.noaa.gov/pub/data/igra/igra2-country-
list.txt"]
```

```
Out[22]=
```

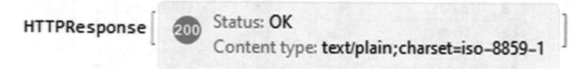

Figure 4-5. *HTTPResponse object of the URL entered*

Now that we know the status, we can proceed to download our data file with URLDownload.

```
In[23]:= URLDownload["https://www1.ncdc.noaa.gov/pub/data/igra/igra2-
country-list.txt"]
Out[23]=
```

File [C:\Users\lb-pc\AppData\Local\Temp\igra2-country-list-7e234436-09c9-45ab-8d0e-673658da5045.txt »]

Figure 4-6. *File object with the locations of the file downloaded*

You should get a file object with location of the file (Figure 4-6), the name, and the extension; in my case, it is in my temporary folder. To open the file in an external viewer, click on the double chevron icon.

Semantic Import

So far, we have seen how to import files of different formats, but there is another tool called SemanticImport that allows us to import files semantically and returns a dataset as a result. Let's see a simple example with the CSV file.

```
In[24]:= SImprt=
SemanticImport["C:\\Users\\My-pc\\Desktop\\Grocery_List.csv"]
Out[24]=
```

id	grocery item	price	sold items	sales per day
1	milk	$4	4	Tue 4 Jun 2019
2	butter	$3	2	Thu 6 Jun 2019
3	garlic	$2	1	Fri 7 Jun 2019
4	apple	$2	4	Sat 1 Jun 2019
5	orange	$3	5	Sat 8 Jun 2019
6	orange juice	$5	2	Sat 8 Jun 2019
7	cheese	$5	2	Thu 6 Jun 2019
8	cookies	$2	5	Sun 9 Jun 2019
9	grapes	$4	3	Fri 21 Jun 2019
10	potatoe	$2	5	Wed 26 Jun 2019

Figure 4-7. *File imported as a dataset with SemanticImport*

As we see in Figure 4-7, when we use semantic import Mathematica, it imports the data in the form of a dataset, and when it does this it recognizes some quantities. These quantities correspond to the magnitude and its units, such as in the case for the elements of the column of price and sales per day. When dealing with quantities, the color of the elements changes, as we see in the dataset, the elements appear differently from the other contents; this is because they are now represented by a semantic type object. Semantic objects include quantities, entities, dates, and geolocation. In other words, they are interpretations made by the freeform interpreter that is related to the Wolfram Alpha Knowledgebase.

In the case of imported data, there are two date type objects, which we saw in the first chapter and quantity type. It should be understood that to work with quantities, we must understand where they come from.

Quantities

The Quantity command converts a magnitude with units to a quantity type, to convert the magnitude with their respective units; the magnitude is entered first, followed by its units in string type. When we do this, Mathematica will display the autocomplete menu as in other occasions. The following example shows it.

```
In[25]:= Quantity[2,"USDollars"]
Out[25]= $ 2
```

Thus, it is transformed into quantity type. When we hover over the result, an ad will be displayed, marking that a result is already a unit. In this case, it is a unit of US dollars. Now, if we check the head of the expression, it will give us the result that it is type quantity.

Note   Quantities are shown in light brown color.

```
In[26]:= Quantity[2,"USDollars"]//Head
Out[26]= Quantity
```

We can also use the inline freeform input, which is in the menu bar: Insert ➤ Inline Freeform Input. This type of input is the input associated with the search engine Wolfram Alpha, so the inline freeform input transforms natural language into Wolfram Language input.

The magnitude and quantity are written inside the box. One of the advantages of this type of input is that natural language can be used. As in the following example, we will write the amount of 77 min, which means 77 minutes. Figure 4-8 shows the input cell of the inline freeform input.

```
In[27]:=
```

Figure 4-8. *Free inline freeform input for quantity of 77 minutes*

Out[27]= 77min

To run the code, clock ENTER, since it gives us a result. Some tabs appear, where
we can click a submenu or a checkmark. If we click on the checkmark, it is to accept the
interpretation made. If we believe that the interpretation is different, we can click on
the other option that is alternate interpretations, and it will show a small pop-up where
it lists different interpretations, as the case may be. Figure 4-9 shows the pop-up for the
example.

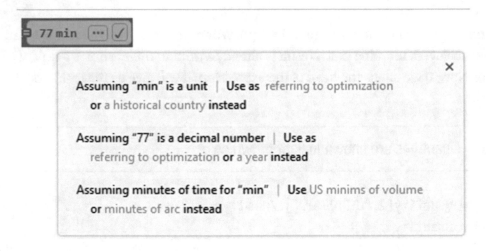

Figure 4-9. *Options for the quantity entered*

Once the interpretation is accepted, the result changes color, and it will be a quantity
type object. And it can be used like any other quantity type object.

When we have quantities, we cannot make operations between numbers, and
quantities are already different types. For these, there are two options to convert the
data to quantities or extract the magnitude of a quantity. To extract the magnitude, the
QuantityMagnitude command is used.

In[28]:= {QuantityMagnitude[77min],Head[%]}
Out[28]= {77,String}

We have already extracted the magnitude, and it is already an integer. In the supposed case of wanting the units, the QuantityUnit command extracts the units.

```
In[29]:= QuantityUnit[77min]
Out[29]= Minutes
```

Datasets with Quantities

There is another aspect to emphasize: To carry out operations, the concept between how to perform arithmetic operations among physical quantities is maintained; otherwise the operation will not be possible, and we will get an error in which the units do not agree. When we carry out an operation between quantities, the result is also of the quantity type, as we will see here.

```
In[30]:=
{77min-77min,77min+77min,77min*77min,77min/77min,77min*3m}
Out[30]= {0min,154min,5929(min)^2,1,231m min}
```

In this example, we see how the results are of type quantity. With the exception of the division, it is already a quotient between the same units. For the last one, it turns out that we have 231 meters per minutes as a result.

Returning to the imported data, we can extract the data from the price column, as shown in Figure 4-10.

```
In[31]:= SImprt[[All,"price"]]
Out[31]=
```

$4	$3	$2	$2	$3
$5	$5	$2	$4	$2

Figure 4-10. *Price column*

If we want to have them in a list, we must use the Normal command.

```
In[32]Normal[%]
Out[32]= {$ 4,$ 3,$ 2,$ 2,$ 3,$ 5,$ 5,$ 2,$ 4,$ 2}
```

147

As we see, the result is the list but in quantity type. It is fair to say that we can do operations with quantities, but if what matters are the magnitudes, we can extract them. Let's see how.

```
In[33]:= QuantityMagnitude[#]&[%]
Out[33]= {4,3,2,2,3,5,5,2,4,2}
```

In this way, we are now working with only the magnitudes.

We can even work with dates and quantities, as we will see in the Figure 4-11, starting by displaying the ID of the products and the date they were sold.

```
In[34]:= SImprt[[All,{"id","sales per day"}]]
Out[34]=
```

id	sales per day
1	Tue 4 Jun 2019
2	Thu 6 Jun 2019
3	Fri 7 Jun 2019
4	Sat 1 Jun 2019
5	Sat 8 Jun 2019
6	Sat 8 Jun 2019
7	Thu 6 Jun 2019
8	Sun 9 Jun 2019
9	Fri 21 Jun 2019
10	Wed 26 Jun 2019

Figure 4-11. *ID and sales per day columns*

Having done this, we can extract the values and work directly with the date object types.

```
In[35]:= Normal[Values[%]]//InputForm
Out[35]//InputForm=
{{1, DateObject[{2019, 6, 4}, "Day", "Gregorian", -5.]},
 {2, DateObject[{2019, 6, 6}, "Day", "Gregorian", -5.]},
 {3, DateObject[{2019, 6, 7}, "Day", "Gregorian", -5.]},
```

```
{4, DateObject[{2019, 6, 1}, "Day", "Gregorian", -5.]},
{5, DateObject[{2019, 6, 8}, "Day", "Gregorian", -5.]},
{6, DateObject[{2019, 6, 8}, "Day", "Gregorian", -5.]},
{7, DateObject[{2019, 6, 6}, "Day", "Gregorian", -5.]},
{8, DateObject[{2019, 6, 9}, "Day", "Gregorian", -5.]},
{9, DateObject[{2019, 6, 21}, "Day", "Gregorian", -5.]},
{10, DateObject[{2019, 6, 26}, "Day", "Gregorian", -5.]}}
```

Note   You should get the DateObject when testing the code instead of the pure word; here, we use the InputForm in order to avoid image conflicts.

Knowing this we can make an association between the ID's of each product and when it was sold, applying the Rule command inside the nested list, followed by creating the associations.

```
In[36]:= Association[Apply[Rule,%,1]]//InputForm
Out[36]//InputForm=
<|1 → DateObject[{2019, 6, 4}, "Day", "Gregorian", -5.],
 2 → DateObject[{2019, 6, 6}, "Day", "Gregorian", -5.],
 3 → DateObject[{2019, 6, 7}, "Day", "Gregorian", -5.],
 4 → DateObject[{2019, 6, 1}, "Day", "Gregorian", -5.],
 5 → DateObject[{2019, 6, 8}, "Day", "Gregorian", -5.],
 6 → DateObject[{2019, 6, 8}, "Day", "Gregorian", -5.],
 7 → DateObject[{2019, 6, 6}, "Day", "Gregorian", -5.],
 8 → DateObject[{2019, 6, 9}, "Day", "Gregorian", -5.],
 9 → DateObject[{2019, 6, 21}, "Day", "Gregorian", -5.],
 10 → DateObject[{2019, 6, 26}, "Day", "Gregorian", -5.]|>
```

To illustrate this, we can create a visualization in a timeline (Figure 4-12), marking the product sold and the date of its sale.

```
In[37]:= TimelinePlot[%]
Out[37]=
```

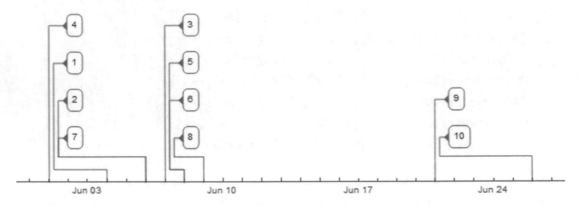

Figure 4-12. *Timeplot*

At this point, we can see the date of every grocery item sold by their ID. When passing the cursor over the number in the timeline, a tooltip pops up showing the exact date.

The idea is that when we use SemanticImport, we can integrate different forms of the Wolfram Language and how we can use this to our advantage when importing our data. With semantic import, it is possible to compare data with other selected data. SemanticImport provides us with tools so that we can work among various types of semantic objects. What is important to observe is that instead of importing common text, we can import currency types, dates, and pretty much any magnitude with the respective unit, as in the previous examples. This allows that data to be associated with different commands within the Wolfram Language.

Costume Imports

Having said all this about semantic import, we can import data and choose how each column in the imported file should be interpreted. However, based on the same idea that we saw earlier, with semantic import we can also choose what data to import (e.g., if it is only one column or several). This is illustrated in Figure 4-13.

```
In[38]:= SemanticImport["C:\\Users\\My-pc\\Desktop\\Grocery_List.csv",
{"Integer","String","Currency","Real", "Date"}]
Out[38]=
```

id	grocery item	price	sold items	sales per day
1	milk	$4	4.0	Tue 4 Jun 2019
2	butter	$3	2.0	Thu 6 Jun 2019
3	garlic	$2	1.0	Fri 7 Jun 2019
4	apple	$2	4.0	Sat 1 Jun 2019
5	orange	$3	5.0	Sat 8 Jun 2019
6	orange juice	$5	2.0	Sat 8 Jun 2019
7	cheese	$5	2.0	Thu 6 Jun 2019
8	cookies	$2	5.0	Sun 9 Jun 2019
9	grapes	$4	3.0	Fri 21 Jun 2019
10	potatoe	$2	5.0	Wed 26 Jun 2019

Figure 4-13. *Dataset with specified data type for columns content*

With this result we observe that we have chosen for the first column to be imported as integers, the second as text, the third as a currency type quantity, the fourth as a real number, and the last as a date object. Having done this, it is possible in the same way that with spreadsheet files we can import certain types of information in list form, either by column or by row. In the following example, we will import rows 1 through 5.

```
In[39]:=
SemanticImport["C:\\Users\\My-pc\\DesktopGrocery_List.
csv",Automatic,"Rows"][[1;;5]]//InputForm
Out[39]//InputForm=
{{1, "milk", Quantity[4, "USDollars"], 4, DateObject[{2019, 6, 4}, "Day",
"Gregorian", -5.]}, {2, "butter", Quantity[3, "USDollars"], 2,
DateObject[{2019, 6, 6}, "Day", Gregorian", -5.]}, {3, "garlic",
Quantity[2, "USDollars"], 1, DateObject[{2019, 6, 7}, "Day", "Gregorian", -5.]},
{4, "apple", Quantity[2, "USDollars"], 4, DateObject[{2019, 6, 1}, "Day",
"Gregorian", -5.]}, {5, "orange", Quantity[3, USDollars"], 5,
DateObject[{2019, 6, 8}, "Day", "Gregorian", -5.]}}
```

As indicated, columns can also be imported, importing from column 1 to 2.

```
In[40]:= SemanticImport[
"C:\\Users\\My-pc\\Desktop\\Grocery_List.csv",Automatic,"Columns"][[1;;2]]
Out[40]= {{1,2,3,4,5,6,7,8,9,10},{milk,butter,garlic,apple,orange,orange
juice,cheese,cookies,grapes,potatoe}}
```

It is necessary to emphasize that if we want to exclude data, then it is recommended to import with the ExcludedLines statement. For example, exclude rows 9 and 10, remembering that the titles are in row 1. This is shown in Figure 4-14.

```
In[41]:= SemanticImport[
"C:\\Users\\lb-pc\\Desktop\\Book Project\\Grocery_List.csv",ExcludedLines
→{{10},{11}}]
Out[41]=
```

id	grocery item	price	sold items	sales per day
1	milk	$4	4	Tue 4 Jun 2019
2	butter	$3	2	Thu 6 Jun 2019
3	garlic	$2	1	Fri 7 Jun 2019
4	apple	$2	4	Sat 1 Jun 2019
5	orange	$3	5	Sat 8 Jun 2019
6	orange juice	$5	2	Sat 8 Jun 2019
7	cheese	$5	2	Thu 6 Jun 2019
8	cookies	$2	5	Sun 9 Jun 2019

Figure 4-14. *Dataset with excluded rows*

Export

Just like with Import, Mathematica supports a lot of format, to view all supported formats type $ExportFormat.

```
In[42]:= Short[$ExportFormats,5]
Out[42]//Short= {3DS,ACO,AIFF,AU,AVI,Base64,Binary,Bit,BMP,BSON,Byte,BYU,BZ
IP2,C,CDF,CDXML,Character16,Character32,Character8,CML,<<148>>,VRML,VTK,WAV
,Wave64,WDX,WebP,WLNet,WMF,WMLF,WXF,X3D,XBM,XHTML,XHTMLMathML,XLS,XLSX,XML,
XYZ,ZIP,ZPR}
```

Exporting data is carried out using the Export command. Export has the form Export["directory path", expr, "format"].

Before starting we need to set a working directory. If not the file will be exported in the default Mathematica working directory. To see the working default directory, use Directory.

```
In[43]:= Directory[]
Out[43]= C:\Users\My-pc\Desktop
```

In my case, the default directory is my Desktop folder.

There are two commands that are key; one is SetDirectory, whose argument is the path of the new working directory, and the other is NotebookDirectory, which is the location of the file.

First let's set the new working directory to export files into our notebook location. By using the notebook directory as argument on SetDirectory, we tell Mathematica that the new working directory will be the location of the notebook in which we are currently working.

```
In[44]:= SetDirectory[NotebookDirectory[]]
Out[44]= C:\Users\My-pc\Desktop
```

Now that we have set a new directory, we can export data created in Mathematica. In the next example, we will export a list of prime numbers from 1 to 10 as a table in a text file and a CSV file. An option applies as well as Import, but if the file extension is added, then it is not compulsory to write the format option.

Note   There is no restriction about whether to assign a name to the list of data or to create the data directly in the export.

```
In[45]:=
Mydata=Table[Prime[i],{i,1,10}];
{Export["New_File.txt",Mydata,"Table"],
Export["New_File.csv",Mydata]}
Out[45]= {New_File.txt,New_File.csv}
```

The output generates the name of the new file exported. An alternative is manually entering the desired location of the file instead of setting a new working directory; I will set my Desktop as the new location.

```
In[46]:=
Export["C:\\Users\\My-pc\\Desktop\\New_File.TSV",Mydata,"TSV"]
Out[46]= C:\Users\My-pc\Desktop\New_File.TSV
```

Now that we have exported the data into a new location, the output is the full path of the new file. If we want to open the file from Mathematica, we can use SystemOpen. This command opens the operating system explorer.

```
In[47]:=
SystemOpen["C:\\Users\\My-pc\\Desktop\\New_File.TSV"]
```

With SystemOpen we can open the notebook directory folder, in case we wanted to open other files inside the notebook directory.

```
In[48]:=
SystemOpen[NotebookDirectory[]]
```

On the other hand, when dealing with tabular data, it be can exported as a spreadsheet. In the next example we will export a tabular data structure and then export them into a spreadsheet format.

To create tabular data, we will use the command Table.

```
In[49]:=
TabD1=Table[i,{i,4}];
TabD2=SetPrecision[Table[i/11,{i,4}],3];
```

Now that we have a set of coordinates, we can export the data to different sheets by typing the reference name of the data into a list of options: {data_sheet 1,data_sheet 2, ...}

```
In[50]:= Export["Tabular_data.xls",{{TabD1},{TabD2}}]
Out[50]= Tabular_data.xls
```

By opening the file with a spreadsheet viewer, you should get that TabD1 is in sheet 1 and TabD2 is in sheet 2.

To customize the name of the sheets, we need to enter the names as a list of rules with the rule operator (➤).

```
In[51]:=
Export["Tabular_data_2.xls",{"Page number 1"→ TabD1,"Page number 2"
→TabD2}]
Out[51]= Tabular_data_2.xls
```

If you open the file, now you should have two sheets with the names we have set.

In addition to this, there is the possibility to add the same data in a single spreadsheet. To do this you only have to enclose the data that we want in the same sheet in curly braces.

```
In[52]:= Export["New_data.xls",Transpose[{TabD1,TabD2}]]
Out[52]= New_data.xls
```

When opening the file, you should have something as shown in the following code.

```
In[53]:= Grid[Transpose[{TabD1,TabD2}]]
Out[53]=
1       0.0909
2       0.182
3       0.273
4       0.364
```

You can even export tables.

```
In[54]:= table1={{"Dog","Wolf"},{"Cat","Leopard"},{"Pigeon","Shark")};
Export["Animal_table.xls",table1]
Out[54]= Animal_table.xls
```

Other Formats

By advancing the topic, it is possible to export the data to simple formats such as TXT, DAT, CSV and CSV. To do this, we only have to put the path of the file where we want it to be exported along with the name of the new file followed by the extension of the desired file. The second argument writes the data to be exported or the variable that contains the data. The third argument is what designates the type of format we want the data to import. Let's look at the following example, where we'll export new data to text and DAT formats. In our case, we only write the name of the file, as this indicates that we want them to be exported to the working directory that we established earlier, which corresponds to the directory of the notebook.

```
In[55]:=
NewD=Table[{i+j,i*j},{i,1,5},{j,1,5}];
{Export["File_text.txt",NewD,"Text"],Export["File_dat.dat",NewD,"Table"]}
```

```
Out[55]= {File_text.txt,File_dat.dat}
```

Here it is advisable to pause for a moment, as we see in the earlier code; we chose the Table format for the DAT file. This is because Table is used so that the exported data becomes an expression in the Wolfram Language. After you have exported, verify that the files have been exported.

Likewise, we can choose the format for a file. For example, instead of type text, we will export in the format of TSV.

```
In[56]:= Export["File_text.txt",NewD,"TSV"]
Out[56]= File_text.txt
```

Similarly, we can export for CSV and TSV files.

```
In[57]:=
{Export["File_csv.csv",NewD,"CSV"],Export["File_tsv.tsv",NewD,"TSV"] }
Out[57]= {File_csv.csv,File_tsv.tsv}
```

There is the possibility to add titles to the columns to the data for when they are exported, either CSV or TSV.

```
In[58]:=
Export["File_csv.csv",NewD,"CSV",TableHeadings→{"column 1","column
2","column 3","column 4","column 5" }]
Out[58]= File_csv.csv
```

With this in mind, it is also possible to define a list of names for the columns, as follows.

```
In[59]:= Labels={"Coordindates 1","Coordinates 2","Coordindates 3",
"Coordinates 4","Coordindates 5"};
Export["File_csv.csv",NewD,"CSV",TableHeadings→Labels]
Out[59]= File_csv.csv
```

In the same way you can export dataset to known formats. As an example, we will use automobile braking distance statistics depending on speed. For this we load the data with the ExampleData command. Inside these we search Statistics, and within this we search CarStoppingDistances.

```
In[60]:= SpData=ExampleData[{"Statistics","CarStoppingDistances"}]
Out[60]= {{4,2},{4,10},{7,4},{7,22},{8,16},{9,10},{10,18},{10,26},{10,34},
{11,17},{11,28},{12,14},{12,20},{12,24},{12,28},{13,26},{13,34},{13,34},
{13,46},{14,26},{14,36},{14,60},{14,80},{15,20},{15,26},{15,54},{16,32},
{16,40},{17,32},{17,40},{17,50},{18,42},{18,56},{18,76},{18,84},{19,36},
{19,46},{19,68},{20,32},{20,48},{20,52},{20,56},{20,64},{22,66},{23,54},
{24,70},{24,92},{24,93},{24,120},{25,85}}
```

To get detail of the dataset on the columns and a brief description, we added Description and ColumnDescriptions.

```
In[61]:= ExampleData[{"Statistics","CarStoppingDistances"},#]&/@{"Descripti
on","ColumnDescriptions"}
Out[61]= {Car stopping distances as a function of speed.,{Speed in miles
per hour.,Stopping distance in feet.}}
```

Continuing the exploration, we see that the first numbers represent the speed in miles per hour and the second numbers represent the distance in feet.

Note   For more information, add Properties as the second argument to ExampleData.

Moving forward in the exercise, we can add the column titles. This will serve to distinguish each type of data when we build the data set (Figure 4-15).

```
In[62]:=
SpDataset=Dataset[SpData,Background→LightBlue][All,<|#1→1,#2→2|>]&["Speed
in miles per hours","Stopping distance in feet"]
Out[62]=
```

Speed in miles per hours	Stopping distance in feet
4	2
4	10
7	4
7	22
8	16
9	10
10	18
10	26
10	34
11	17
11	28
12	14
12	20
12	24
12	28
13	26
13	34
13	34
13	46
14	26

⌐ ∧ rows 1–20 of 50 ∨ ⌐

Figure 4-15. *CarStoppingDistances dataset*

With this we have concluded the creation of the dataset. Now this data, together with their respective column titles, can be exported to a CSV format.

```
In[63]:= Export["Dataset_csv.csv",SpDataset,"CSV"]
Out[63]= Dataset_csv.csv
```

If the export is successful, then you should have a CSV file in the correct format. For the case of a TSV file, see the following form.

```
In[64]:= Export["Dataset_tsv.tsv",SpDataset,"TSV"]
Out[64]= Dataset_tsv.tsv
```

XLS and XLSX Formats

It is worth distinguishing that to export datasets to spreadsheet formats such as XLS or XLSX, we should work the dataset as a list, since trying to export the dataset directly would result in exporting associations in a single cell, and we are not interested in that. Regarding the second point, since we have our dataset, to extract the values we use the Normal command, which converts the dataset into a normal expression followed by extracting the values from the braces with Values.

```
In[65]:= Values@Normal@SpDataset
Out[65]= {{4,2},{4,10},{7,4},{7,22},{8,16},{9,10},{10,18},{10,26},{10,34},
{11,17},{11,28},{12,14},{12,20},{12,24},{12,28},{13,26},{13,34},{13,34},
{13,46},{14,26},{14,36},{14,60},{14,80},{15,20},{15,26},{15,54},{16,32},
{16,40},{17,32},{17,40},{17,50},{18,42},{18,56},{18,76},{18,84},{19,36},
{19,46},{19,68},{20,32},{20,48},{20,52},{20,56},{20,64},{22,66},{23,54},
{24,70},{24,92},{24,93},{24,120},{25,85}}
```

Now that we have the data, we can move on to adding the column titles and then export the extracted data from the dataset.

```
In[66]:=
ColTitles={"Speed in miles per hours","Stopping distance in feet"};
```

To attach the two lists, we will use Prepend. And we'll assign the name ExprtData to new values.

```
In[67]:= Short[ExprtData=Prepend[%%,ColTitles],1]
Out[67]//Short= {{Speed in miles per hours,Stopping distance in feet},
{4,2},{4,10},{7,4},{7,22},{8,16},<<39>>,{23,54},{24,70},{24,92},{24,93},
{24,120},{25,85}}
```

It should be noted that we do not define variables to put together the data list and titles. We use percentage notation to simplify the code. Now that we have our complete data, we can proceed to export it to an XLS or XLSX format.

```
In[68]:= Export["Stopping_distance_Dataset.xlsx",ExprtData,"XLSX"]
Out[68]= Stopping_distance_Dataset.xlsx
```

If we verify the file, we should have something like the dataset created earlier.

JSON Formats

Leaving the aforementioned aside, it is also possible to export information to formats such as JSON. In the following example, we will create a JSON structure from an association.

```
In[69]:= Association@
{
"Name"→"Ellis",
"Date of birth"→"1990,01,04",
"Height"→"180 cm",
"Favorite color"→"Red",
"Hobbies"→"Soccer, Pc gaming, Board games",
"Social netwoks"→"Twitter, Facebook"
};
Export["File_json.json",%,"JSON"]
Out[69]= File_json.json
```

If you open the new JSON file, you will see that it has a structure corresponding to a JSON file. For the case where we have a nested list, it is the same process, although you can also use the "Rawjson" format when exporting. The idea is that we can export data to JSON formats, from associations, as we have seen; the braces and values of an association can be any expression. This leads us to say that more associations can be added, and these can be exported. The important thing to note is that given the nature of the JSON format of containing braces and values in pairs, it is possible to export data in JSON format from associations. Examining the case for when we have a dataset (Figure 4-16), proceed as noted here.

```
In[70]:= Association@
{
"Name"→"Ellis",
"Date of birth"→DateObject[{1990,01,04}],
"Height"→Quantity[180,"Centimeters"],
"Favorite color"→"Red",
```

```
"Hobbies"→"Soccer, Pc gaming, Board games",
"Social netwoks"→"Twitter, Facebook"
};
User=Dataset[%]
Out[70]=
```

Name	Ellis
Date of birth	Thu 4 Jan 1990
Height	180 cm
Favorite color	Red
Hobbies	Soccer, Pc gaming, Board games
Social netwoks	Twitter, Facebook

Figure 4-16. *JSON file dataset*

As we see the dataset is built, but in some cases the dataset may contain quantities or other semantic objects, as in this case, the date and height. So, to export them would be the same way as before but using the JSON option format, not Rawjson, since this does not allow exporting dataset objects. If we want to use Rawjson, we must convert the semantic objects to string or numbers.

```
In[71]:= Export["Dataset_json.json",User,"JSON"]
Out[71]= Dataset_json.json
```

If we have a dataset of repeated keys, we can export it to the JSON format (Figure 4-17).

```
In[72]:=
Assoc1=<|"Log in Date"→DateObject[{2020,06,29}],"User
ID"→123,"Status"→"Active"|>;
Assoc2=<|"Log in Date"→DateObject[{2020,06,28}],"User
ID"→122,"Status"→"Not Active"|>;
Dataset[{Assoc1,Assoc2}]
Export["Dataset2_json.json",%,"JSON"]
Out[72]=
```

Log in Date	User ID	Status
Mon 29 Jun 2020	123	Active
Sun 28 Jun 2020	122	Not Active

Figure 4-17. *Dataset*

```
Out[72]= Dataset2_json.json
```

To be precise, you can export shapes where the dataset contains complex structures such as an association of associations. Let's look at the following example, where we first build a dataset (Figure 4-18).

```
In[73]:= Assoc3="Player A"→Association["Date"→DateObject[{2020,06,29}],
"User ID"→123,"Status"→"Active"];Assoc4="Player B"→Association["Date"→
DateObject[{2020,06,28}],"User ID"→122,"Status"→"Not Active"];
Dataset[{<|Assoc3,Assoc4|>}]
Out[73]=
```

	Date	User ID	Status
Player A	Mon 29 Jun 2020	123	Active
Player B	Sun 28 Jun 2020	122	Not Active

Figure 4-18. *Tagged dataset*

Subsequently we proceed to export the dataset.

```
In[74]:= Export["Dataset3_json.json",%,"JSON"]
Out[74]= Dataset3_json.json
```

To better understand how to export in JSON format. When we export information such as a rule list or a single association, the structure of the content in JSON file that is exported will be through a collection of pairs between braces and values. On the contrary, when we have ordered structures, such as an association of lists and association of associations, the structure of the content in the JSON file will be as an ordered array within the array of the collections of associated pairs between braces and values. Quite the opposite, however, when we export a nested list, it will already be in the form of sorted arrays. To clarify this, the reader is invited to observe how a list of rules is exported through the following code.

```
In[75]:= Rules={"apple"→3,"car"→"3","2"→2};
Export["Rules.json",Rules,"JSON"]
Out[75]= Rules.json
```

In addition, for a nested list or list of lists.

```
In[76]:= Arry=Array[{#1,#2}&,{4,4}]
Export["Array.json",Arry,"JSON"]
Out[76]= {{{1,1},{1,2},{1,3},{1,4}},{{2,1},{2,2},{2,3},{2,4}},
{{3,1},{3,2},{3,3},{3,4}},{{4,1},{4,2},{4,3},{4,4}}}
Out[76]= Array.json
```

If the created file is observed, it must contain an array of arrays inside the JSON file.

Content File Objects

It should be concluded that for all the files that are exported, we can create a content object that can show us properties of the created files. This is carried out with the ContentObject function, which gives us content from a file. To do this, let's take the example of the association to create a JSON file.

```
In[77]:=
Association@
{
"Name"→"Ellis",
"Date of birth"→DateObject[{1990,01,04}],
"Height"→Quantity[180,"Centimeters"],
"Favorite color"→"Red",
"Hobbies"→"Soccer, Pc gaming, Board games",
"Social netwoks"→"Twitter, Facebook"
};
User=Dataset[%];
JsonFile=Export["Dataset_json_2.json",User,"JSON"];
```

Now we need to get the path where the file is located with AbsoluteFileName.

```
In[78]:= AbsoluteFileName[JsonFile]
Out[78]= C:\Users\My-pc\Desltop\Dataset_json_2.json
```

We now use File to create the file object type representation.

```
In[79]:= File[%]//InputForm
Out[79]//InputForm=
File["C:\\Users\\My-pc\\Desktop\\Dataset_json_2.json"]
```

Then ContentObject is applied to the file object.

```
In[80]:= ContentObject[%]
Out[80]=
```

Figure 4-19. *ContentObject for the JSON files created*

A content type object will appear (Figure 4-19). If we press the + icon, it will provide us with properties of the exported file, such as name, size, creation dates, and file localization. You can access the properties programmatically using the following form.

```
In[81]:= ContentObject[%%]["Properties"]
Out[81]= {Location,FileName,ModificationDate,CreationDate,FileByteCount,
FileExtension,Title,Plaintext}
```

This can be applied to other exported files.

Searching Files with Wolfram Language

With the Wolfram Language we can look at the locations of file or files.

To see the path of the notebook directory, the command NotebookDirectory is used. It will show the full directory containing the notebook in which you are currently working.

```
In[82]:= NotebookDirectory[]
Out[82]= C:\Users\My-pc\Desktop\
```

Now SetDirectory is used to set a working directory as the current directory. You can enter the path of the desired directory and establish it as the working directory. However, for now we will set the notebook directory as the new working directory.

```
In[83]:= SetDirectory[NotebookDirectory[]]
Out[83]= C:\Users\My-pc\Desktop
```

With this new directory set, we can now proceed to located files in the new directory, which is the notebook location. The command FileNames lets us explore for files that are in the working directory, which in this case is the notebook's directory because we set it up in the previous code.

```
In[84]:= FileNames[]
Out[84]= {Color_table.txt,Grocery_List.csv,Hello_World,Hello_World.
txt,import export.nb,weather.csv}
```

FileNames will show all types of files available in the directory. If we have a lot of files in the directory, we can search for a particular file by using FindFile and entering the name of the file as a string. And the full path of the file will be displayed.

```
In[85]:= FindFile["Color_table.txt"]
Out[85]= C:\Users\My-pc\Desktop\Color_table.txt
```

File extension can be searched too.

```
In[86]:= FileNames["*.txt"]
Out[86]= {Color_table.txt,Hello_World.txt}
```

Note   Other types of File commands exist; to look for more commands associated with the name file, enter ??File*.

Remember this is the case when we set the working directory as the notebook directory. If we have not set a directory previously, Mathematica will search the default directories of your machine, which are the ones shown entering $Path.

CHAPTER 5

Data Visualization

In this chapter we will see more depth in terms of data visualization, where we will see the different ways of representing data visually, with the use of different commands, and create a range of different types of graphs. We will also see how to customize plots and use predefined plot themes.

Basic Visualization

Data visualization is key for understanding information about our data. Visual tools such as 2D plots, contour plots, 3D plots, time series, etc. provide a handy form to view and understand trends and patterns of the data. One of the things about Wolfram Language is that it contains commands that enable us to plot graphs in a simple form. Now, we will better learn how plotting works. Mathematica treats every plot as a graphic object, that is because every graphic is created of primitive elements (points, lines, polygons, geometric figures, etc.), directives (style, shape, size, width, blurriness, etc.), and options (visual modifications, styles, frames, aspects, text, etc.). However, we will only center on the area of 2D and 3D plots.

2D Plots

Simple 2D plots over a specified range are fairly simple to create, like we saw in Chapter 1 with the function Plot. The Wolfram Language gives you accurate control over your plots; for example, you can define the range of your plot, as well as many options. For instance, we can add a title to the next plot, which is a LogPlot, a function in a logarithm scale (Figure 5-1).

```
In[1]:= LogPlot[Log[x]/x,{x,1,20},PlotLabel→"New Log plot"]
Out[1]=
```

© Jalil Villalobos Alva 2021
J. Villalobos Alva, *Beginning Mathematica and Wolfram for Data Science*,
https://doi.org/10.1007/978-1-4842-6594-9_5

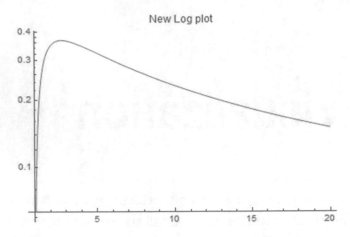

Figure 5-1. *LogPlot*

In Figure 5-1, we can now see that a title has been added.

When plotting points over an interval, the default plot range to show is produced automatically by Mathematica, but with PlotRange we can override the option and enter a desired range (Figure 5-2).

```
In[2]:= LogPlot[x+(6/x),{x,1,20},PlotLabel→"New Log
plot",PlotRange→{0,14}]
Out[2]=
```

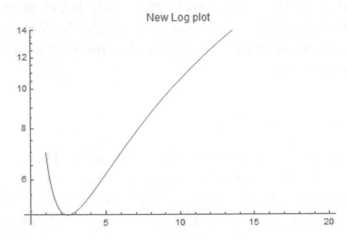

Figure 5-2. *LogPlot of x+ (6/x), with custom range*

By selecting All in PlotRange, the y axis increases. Alternatively, we can choose the limits by entering them in the form {y min, y max}. Now sometimes a graphic may not pass through a desired set of coordinates; to force this, AxesOrigin is used (Figure 5-3). Intersections are written in the form {x,y}, where the coordinates denote the x and y origin point

```
In[3]:= Plot[Abs[x],{x,-2,2},AxesOrigin→{0,2}]
Out[3]=
```

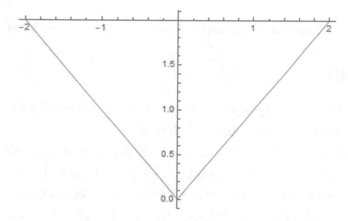

Figure 5-3. *Absolute value of x on origin 0, 2*

To control the aspects by means of their height and width, AspectRatio is used. This option allows us to specify how big or small a graphic can be, the ratio is calculated by height to width (h/w). An alternative to control the aspect of the graphics is ImageSize. ImageSize allows you to modify the size of the graphic both options are shown in Figure 5-4.

```
In[4]:=
GraphicsRow[{Plot[Cos[x],{x,0,2\[Pi]},ImageSize→Small],
Plot[Cos[x],{x,0,2\[Pi]},AspectRatio→0.5]}]
Out[4]=
```

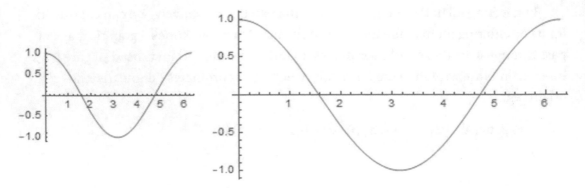

Figure 5-4. *First graphic with ImageSize; second with AspectRatio*

Plotting Data

When plotting graphs, a set of points can be represented in a plot. Data can be plotted with different commands, depending on their purpose.

To plot a list of coordinates, ListPlot is used, the arguments of the plot are represented as x,y coordinates ($\{x_1,y_1\}$, $\{x_2,y_2\}$ ...). We can create a list of values and pass them as the arguments. In the next example we create a table of values to resemble a hyperbolic cosine, with one step between each point (Figure 5-5).

```
In[5]:= ListPlot[Table[Cosh[i Degree],{i,1,20}]]
Out[5]=
```

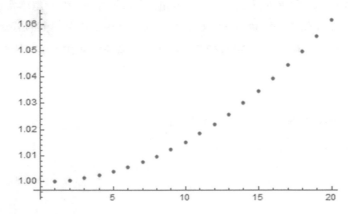

Figure 5-5. *Hyperbolic cosine plot*

In this case we only generate points in the form of $\{1, y_1\}, \{2, y_2\}$, but we can also plot our x values and y values. For this case we will generate the x points with Table and then thread each element of x to a y element and plot (Figure 5-6) the new set of coordinates.

```
In[6]:= xcoor=Table[i,{i,1,5}];
ycoor={12,5,35,20,55};
Coordinates=Thread[{xcoor,ycoor}];
ListPlot[Coordinates]
Out[6]=
```

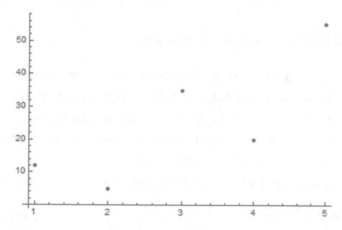

Figure 5-6. *ListPlot of x and y coordinates*

Another useful command is ListLinePlot, which plots points through points by joining them with a line. ListLinePlot (Figure 5-7) can also plot a set of predefined coordinates. We can show how many points to display in order to understand how the construction of the plot is made with the Mesh option.

```
In[7]:= ListLinePlot[Coordinates,Mesh→20]
Out[7]=
```

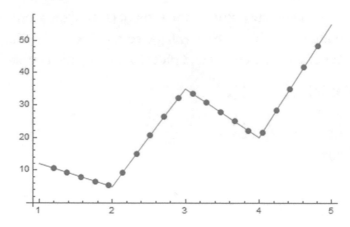

Figure 5-7. *ListLinePlot with mesh option set to 20*

A plot can be represented by different colors and markers. Colors and markers are convenient to distinguish among different plots. To introduce markers, enter the option PlotMarkers followed by the markers symbol. Markers can be special characters or even letters; use the special character pallet for a complete list of symbols and characters. By default, different sets are colored differently, but to choose a specific color use PlotStyle. With PlotStyle the thickness of a line can be changed too, as shown in Figure 5-8.

```
In[8]:= ListLinePlot[{Table[Cos[i ],{i,0,2\[Pi],0.2}],Table[Sin[i],
{i,0,2\[Pi],0.2}]},PlotMarkers→{"\[CloverLeaf]","\[FilledDownTriangle]"},
PlotStyle→ {Green,Black}]
Out[8]=
```

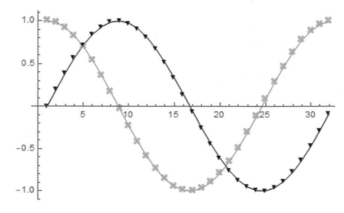

Figure 5-8. *Plots with different marker points*

Another general option is Ticks. With this option we can modify the indicators on the axes for both x and y. For example, in Figure 5-9 the plot ticks are marked on the x-axis, the ticks are -1 and 1. And the y axis, is set to automatic (Figure 5-9).

```
In[9]:= Plot[x^3,{x,-5,5},Ticks→{{-1,0,1}, Automatic}]
Out[9]=
```

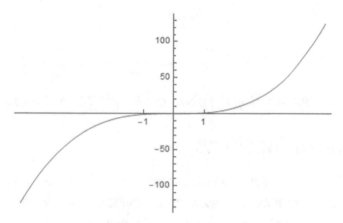

Figure 5-9. *Plots with ticks marked on -1 and 1 for the x axis*

Additionally plots containing dates can be displayed with DateListPlot. The DateListPlot has the next form, DateListPlot[{v1,v2, ... }, "date specification"]. With DateListPlot, the x axis is converted into a timeline and the y axis correspond to the values (v1.v2, ...). The Figure 5-10 shows a DateListPlot, starting at June and finishing in November.

```
In[10]:=
data1=Table[Power[i,2],{i,0,5}];
data2=Table[Power[i,3],{i,0,5}];
DateListPlot[{data1,data2},{2006,06}]
Out[10]=
```

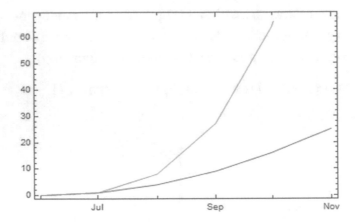

Figure 5-10. *Date plot, starting the plot from June 2006 to November 2006*

Plotting Defined Functions

Just as we can define custom functions, these functions can be plotted (Figure 5-11). User functions can also be used as arguments for plotting commands. Functions can have a single or multiple variables, as we will see with 3D plots.

```
In[11]:= F[x_]:=Exp[x];
Plot[F[x],{x,-10,10}]
Out[11]=
```

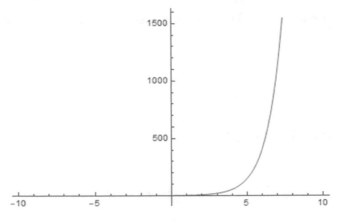

Figure 5-11. *User-defined function for Exp of x*

Also, multiple defined functions are supported. When multiple plots are in the same graphic, each plot is colored differently (Figure 5-12).

174

```
In[12]:= X[x_]:=x; Y[y_]:=-Sqrt[y];Z[z_]:=1/z;
Plot[{X[x],Y[x],Z[x]},{x,-10,10}]
Out[12]=
```

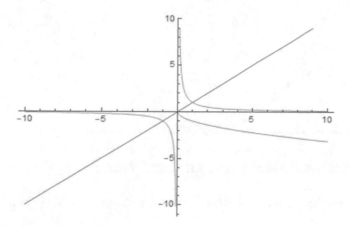

Figure 5-12. *Multiple plots*

Customizing Plots

The Wolfram Language lets users customize plots based on their needs, like adding text, changing color style, adding fill, presenting on tabular frameworks, etc. Many commands used in the 2D plots are also preserved in 3D plots. Depending on the graphical representation, options can vary between commands.

Adding Text to Charts

Adding text to charts can make a chart more informative, like markers and the range of values. Many other elements can be added too.

As seen before, PlotLabel adds a title to our chart. In addition to this option, there is AxesLabel and PlotLegends. The first allows us to add labels to our axes in the form {"x_label", "y_label"}; the second allows us to add text related to each expression within the graph (Figure 5-13).

```
In[13]:= Plot[{Abs[x],x^2},{x,-2,2},AxesLabel→{"x","y"},PlotLegends→
"Expressions"]
Out[13]=
```

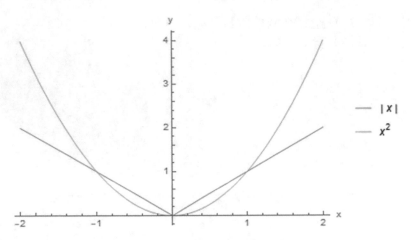

Figure 5-13. *Plots with labeled axes and functions*

We can use Labeled to add costume text expressions on plots (Figure 5-14).

```
In[14]:= Labeled[Plot[x^2,{x,-2,2}],"f(x) = x²",Left]
Out[14]=
```

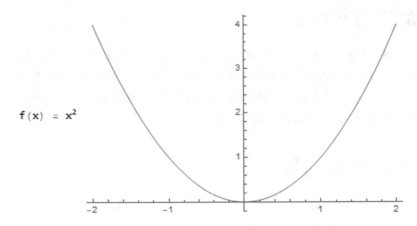

Figure 5-14. *Label placed on the left side on the graphic*

Even with the Labeled command, Tooltips can be constructed. Tooltips displays a label tooltip for any expression (Figure 5-15). Tooltips are displayed when the mouse pointer is passed over the tooltip expression. The difference between Tooltips and PlotLegends is that PlotLegends is an option and not a command.

```
In[15]:= Tooltip[{Plot[x^2,{x,-2,2}]}]
Out[15]= {}
```

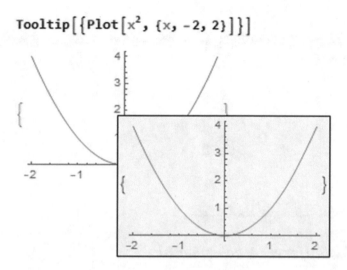

Figure 5-15. *Tooltip created for the plot expression*

When we hover over the entire graph, it will show us the tooltip of the entire graph since we specify it. But we can do it just for the expression of the function (Figure 5-16).

```
In[16]:= Plot[Tooltip[x^2],{x,-2,2},ImageSize→200]
Out[16]=
```

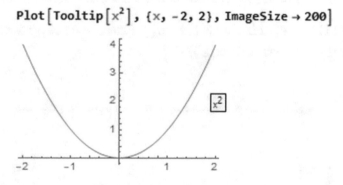

Figure 5-16. *Tooltip for the curve expression*

If we hover over the curve, it will show us the tooltip of x^2; this function is not only limited to Plot—it also works with the other types of plots. We can add what the tooltip style should look like with the ToolTipStyle option (Figure 5-17).

```
In[17]:= ListPlot[Tooltip[Range[10],TooltipStyle → {Bold,Red,Background
→LightBlue}],ImageSize-> 250]
Out[17]=
```

```
ListPlot[
  Tooltip[Range[10], TooltipStyle → {Bold, Red, Background → LightBlue}],
  ImageSize → 250]
```

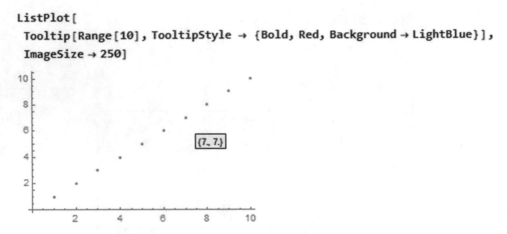

Figure 5-17. *Tooltip for every point plotted*

If we move the cursor to the points, we get the coordinates of the points written in red and the background of the tooltip in light blue.

Frame and Grids

Plots can be framed and gridded. The Frame option is used, and to add labels to the frame use FrameLabel, which receives instructions like AxesLabel (Figure 5-18).

```
In[18]:= ListPlot[Table[Prime[i],{i,1,10}],Frame→ True,FrameLabel
→{"X Framed Axis ","Y Framed Axis"}]
Out[18]=
```

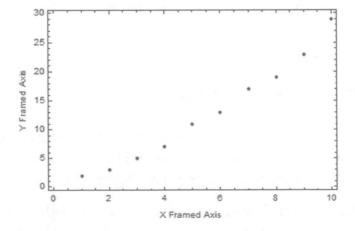

Figure 5-18. *Framed ListPlot*

To add a grid (Figure 5-19), use the GridLines option.

```
In[19]:= ListPlot[Table[Prime[i],{i,1,10}],GridLines→Automatic,AxesLabel→
{"X Framed Axis ","Y Framed Axis"}]
Out[19]=
```

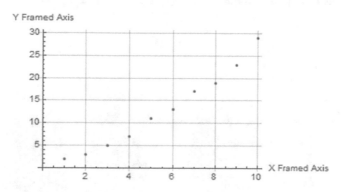

Figure 5-19. *Gridded plot*

To modify the grid style, use the GridLinesStyle option, which can have a particular thickness with the use of Directive (Figure 5-20).

```
In[20]:= ListPlot[Table[Prime[i],{i,1,10}],GridLines→Automatic,GridLines
Style→Directive[Thickness[0.0002],LightRed]]
Out[20]=
```

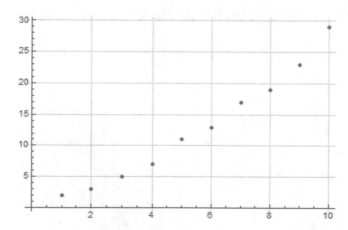

Figure 5-20. *GridLines colored in light red*

Filled Plots

Plots can be filled in various forms—for example, between the x axis, from the bottom and top of a curve (Figure 5-21).

```
In[21]:= ListLinePlot[Table[Mod[i,2],{i,0,5}],Filling→Bottom]
Out[21]=
```

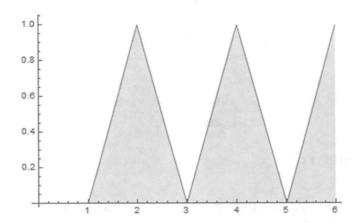

Figure 5-21. *Filled plot from plotted points to the bottom of the axis*

They can also be filled by a specified region between curves, by introducing Filling → {"1st curve" → {"2nd curve"},"2nd curve" → {"3rd curve"}, as shown in Figure 5-22.

```
In[22]:= Plot[{x^2,x^3,x^4},{x,0,5},Filling→{1→{2},2→{3}}]
Out[22]=
```

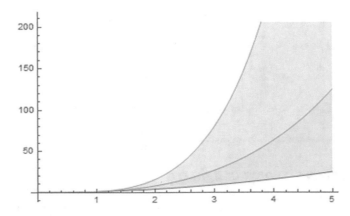

Figure 5-22. *Filled plots*

180

Combining Plots

To display overlap graphics, there are ways to display the graphs even if they are not of the same type. In the next example, we will be assigning names to plots, without showing the result of each one, and finally showing the three graphs. The Show command shows previously defined plots; the arguments of show are graphic objects followed by options. This is an alternative instead of doing multiple listable subplots.

```
In[23]:= Plot1=Plot[x,{x,0,10},PlotStyle→Red];Plot2=Plot[Cos[x],{x,0,10},
PlotStyle→Black];Plot3=ListPlot[Table[Sin[i]+1,{i,1,10}],PlotStyle→Brown];
Show[Plot1,Plot2,Plot3,PlotRange→Automatic]
Out[23]=
```

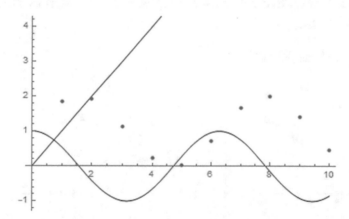

Figure 5-23. *Combined plots shown in the same graphic*

As we see in Figure 5-23, Show changes the appearance of the graphics; the order in which they are entered is preserved when they are displayed. Although it is possible to make the graphics within Show, if we want, we can add colors within the Plot command to distinguish the different graphs (Figure 5-24).

```
In[24]:= Show[Plot[Cos[x],{x,0,10},PlotStyle→Orange],Plot[Sin[x],{x,0,10},
PlotStyle→Purple],PlotRange→Automatic]
Out[24]=
```

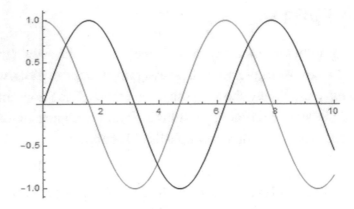

Figure 5-24. *Cosine and Sine plot in the same graphic*

There are several ways to create a list of graphs. We can assign variables to graphs and deploy them as a list.

In[25]:= {Plot1,Plot2,Plot3}
Out[25]=

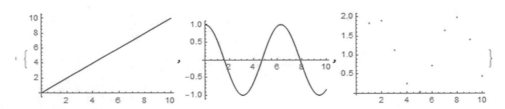

Figure 5-25. *List of three different plots*

As we see in Figure 5-25, these three graphs are separated by commas, since it is a list.

Multiple Plots

Multiple plots can be shown in a single output cell. To do this, use the Row command; this command gives the possibility to display the graphs in horizontal form with each graph on one side of the other (Figure 5-26). However, Row works generally to display expressions in row form, not necessarily just graphs.

In[26]:= Row[{Plot1,Plot2,Plot3}]
Out[26]=

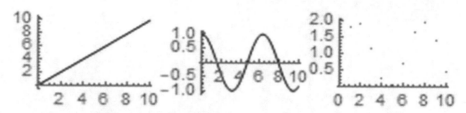

Figure 5-26. *Plots expressed as a row*

By entering a second argument for Row (Figure 5-27), we have the option to add a separator between the graphs.

```
In[27]:= Row[{Plot1,Plot2,Plot3},"**--**"]
Out[27]=
```

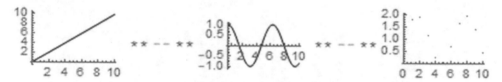

Figure 5-27. *Separator (**--**) added between each plot*

As an alternative to this, there is Column, which acts similarly to Row, with the difference of displaying expressions or graphs in column form (Figure 5-28).

```
In[28]:= Column[{Plot1,Plot2,Plot3}]
Out[28]=
```

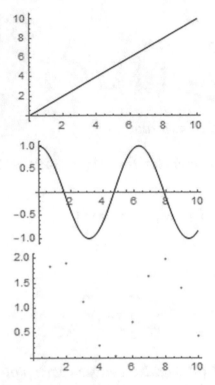

Figure 5-28. *Graphics expressed as a column*

If we look at the following example, it is possible to add frames over the entire chart (Figure 5-29); this is for both Column and Row.

```
In[29]:= {Column[{Plot1,Plot2,Plot3},Frame→True],Row[{Plot1,Plot2,Plot3},
Frame→True,FrameMargins→Medium]}
Out[29]=
```

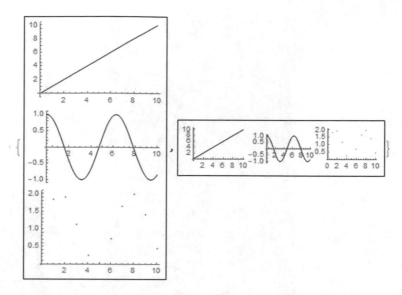

Figure 5-29. *Exhibit of column and row expression for the three plots*

Coloring Plot Grids

With Column and Row, it is possible to customize the graphs in the form of how we want to show them. There are various ways of changing the color of the frame, adding shading to the graphs (Figure 5-30).

```
In[30]:= Column[{Plot1,Plot2,Plot3},Frame→True,Background→LightCyan,
FrameStyle→Directive[Black,Dashed],Dividers→All]
Out[30]=
```

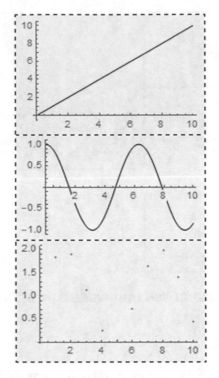

Figure 5-30. *Column graphics with multiple features*

Depending on whether we are using Row or Column, some options will be available. With Column there is the option of dividers; in Row there is not such an option, but it is done via a separator as we saw earlier. With the use of Table, it is possible to create different shapes on the graphs, either by color, frames, etc., as shown in Figure 5-31.

```
In[31]:= Table[Row[{Plot1,Plot2,Plot3},Frame→True,FrameStyle→Opts],{Opts,
{Thick,Dashed,Dotted}}]
Out[31]=
```

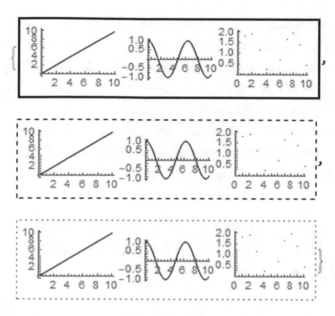

Figure 5-31. *Table of multiple features implemented with the Row command*

After this situation, let us address the alternative that exists with the use of GraphicsRow and GraphicsColumn. Around these commands, there are options also in the image size (Figure 5-32).

```
In[32]:= GraphicsRow[{Plot1,Plot2,Plot3},ImageSize→Medium]
GraphicsColumn[{Plot1,Plot2,Plot3},ImageSize→Small]
Out[32]=
```

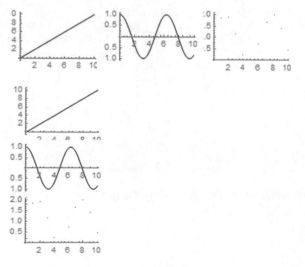

Figure 5-32. *GraphicsRow vs GraphicsColumn*

187

GraphicsRow and GraphicsColumn are commands with specific shapes for constructing graphics, whether polygons, lines, dots, and so on. In addition, with Row and Column, the graphs are independent of each other. With GraphicsRow or GraphicsColumn, if we select the graph, it will be as a unique image that will contain (in this case) the three plots that we have made.

There is still another useful command, which shows us the graphs as a network, taking up the point stated earlier—if we select the graph, this will be a unique image. In the following example, we will add another chart to better illustrate why it's useful to use GraphicsGrid.

```
In[33]:= Plot4=LogLogPlot[Cos[x],{x,0,10},PlotStyle→Yellow];
GraphicsGrid[{{Plot1, Plot2},{Plot3,Plot4}},Frame→All, FrameStyle→Purple,
Background→LightCyan]
Out[33]=
```

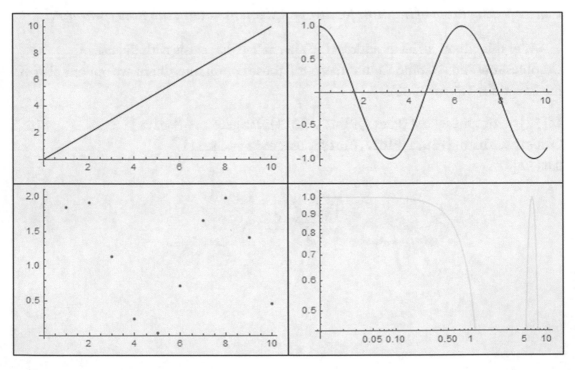

Figure 5-33. *GraphicsGrid showing four different plots*

As we see in Figure 5-33, this shape can help us to compactly visualize four graphs at once. Without a doubt, the graphs do not have to be so simple. The different options that we have seen throughout this chapter can also be added, such as having titles; labels on the axes add grid lines and colors, among many others, as shown in the following example.

```
In[34]:= NewPlot1=Plot[x,{x,0,10},PlotStyle→{Purple,Thick},PlotLabel→"X"];
NewPlot2=Plot[Cos[x],{x,0,10},GridLines→{{-1,0,1}, {-1,0,1}},GridLinesStyl
e→Directive[Dotted, Blue],PlotLabel→"Cos[x]",ColorFunction→"Rainbow"];
NewPlot3=ListPlot[Table[Sin[i]+1,{i,1,10}],Frame→True,FrameLabel→{Style
["X",Bold],Style["Y",Bold]},PlotStyle→Red,PlotMarkers→"X",PlotLabel→"2D
Scatter Plot"];
NewPlot4=LogLogPlot[Cos[x],{x,0,9},Filling→Axis,ColorFunction→"BlueGreenY
ellow",PlotRange→{0,1},PlotLabel→"Log Log Plot"];
```

Now that we have the new plots, we can compare them by putting them as a nested list in GraphicsGrid (Figure 5-34).

```
In[35]:= Labeled[GraphicsGrid[{{NewPlot1,NewPlot2},{NewPlot3,NewPlot4}},
Frame→All,Background→White,Spacings→1],Style["Multiple Plots Box",20,
Italic],Top,Frame→True,Background→LightYellow]
Out[35]=
```

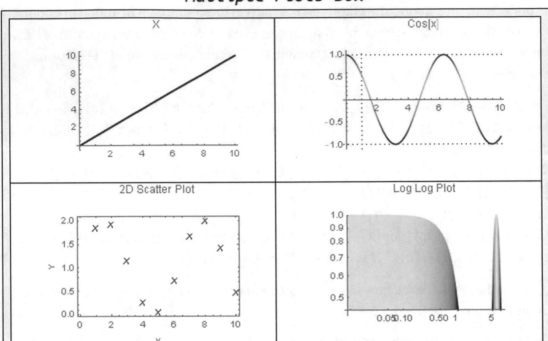

Figure 5-34. *Grid of multiple plots*

I would like to add that this is not restricted to displaying graphs in two dimensions but also applies to graphs with three dimensions and other types of charts.

Colors Palette

If we are interested in more colors, there is a gamma of various types of colors in Mathematica. For this, go to the menu in Palettes → Color Schemes. This will show us the color palette, as shown in Figure 5-35.

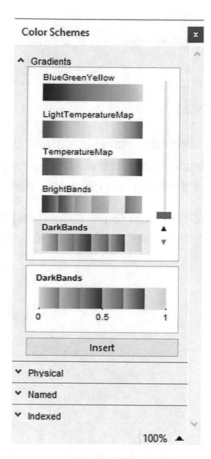

Figure 5-35. *Colors palette*

The tabs that appear are of the colors associated with the different classes. To defer through the colors in the tabs, use the arrows and the different names of the colors and their color or gradient will be displayed. If we want to introduce colors that are not as reserved words, then we use the insert button. For example, we go to the Gradient tab, and then click the insert button. This will insert the function with the chosen color into the notebook.

To illustrate let us look at the following example. Select the Color BrownCyanTones and insert it with the button and evaluate the expression and get the result of the ColorDataFunction (Figure 5-36).

```
In[36]:= ColorData["BrownCyanTones"]
Out[36]=
```

Figure 5-36. *ColorData object*

This gives us a color data object that shows us the name, color type, class, and domain. Gradient colors are intricate in text and work best with the ColorFunction function. So now that we know the name, we can assign it as color (Figure 5-37).

```
In[37]:= Plot[x,{x,0,10},ColorFunction→ ColorData["BrownCyanTones"]]
Out[37]=
```

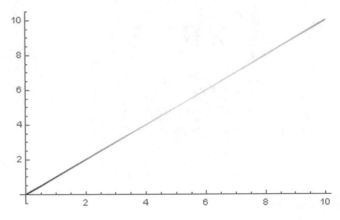

Figure 5-37. *Gradient color of straight line x*

Note   Plain colors are located in the Named tab of the palette.

3D Plots

Mathematica has the capability to perform various types of 3D graphics; many of them are simple. 3D functions are displayed as surfaces in space. Figure 5-38 presents example (Figure 5-38).

```
In[38]:= Plot3D[Sinc[x*8+y^2],{x,-1,2},{y,-1,3},ImageSize→Medium,PlotPoin
ts→20]
Out[38]=
```

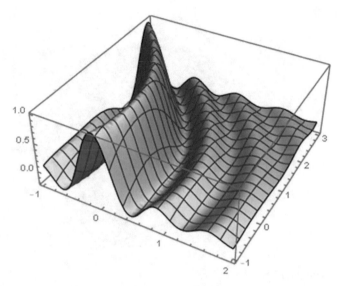

Figure 5-38. *3D plot figure*

Mathematica gives us the functionality to observe the graph by moving with the cursor. Hovering over the chart will change to rotating arrows. This means that we can move the chart to observe it from different points. One last observation is that when you press the CTRL or ALT key, you can make a magnification of the chart keeping its position fixed.

Note that 3D graphs can be manipulated by the cursor, so that we can visualize the graph of angle spreads. Common standard Mathematica displays the graph as a mesh, which can be modified with the Mesh option, as we saw earlier, or by adding more points to evaluate with the PlotPoints option. This increases the number of points in both directions in both x and y. It also serves to improve the quality of the chart.

Customizing 3D Plots

3D graphics can also be customized as 2D graphics (Figure 5-39) as labels to axes, color, grids, etc. Figure 5-39 shows a 3D plot with the options of AxesLabel, ColorFunction and FaceGrids

```
In[39]:=Plot3D[Sin[4(x^2+y^2)]/0.5,{x,-0.8,0.8},{y,-0.8,0.8},AxesLabel
→{"X axis","Y axis","Z axis"},ColorFunction→"Rainbow",FaceGrids→All]
Out[39]=
```

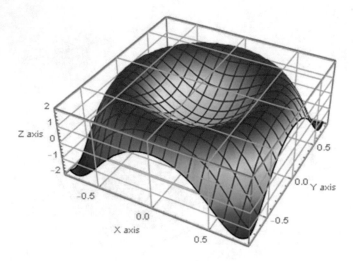

Figure 5-39. *Gridded 3D plot*

Table 5-1 shows general options for 3D graphics.

Table 5-1. *Plot Options Table*

Option	Instructions
AspectRatio	Height/width ratio.
AxesLabel	Add text to axes
PlotStyle	Color, opacity, thickness, etc.
PlotRange	Range of values
PlotLabel	Plot title
Background	Background color

Customization of graphics depend on how you plan to exhibit them. There is no limit on how graphics are presented. In the next example, we plot a 3D function and color the background in light yellow (Figure 5-40).

```
In[40]:= Plot3D[Sin[0.9(x^2+y^2)]/0.5,{x,-1,1},{y,-1,1},AxesLabel
→{"X axis","Y axis","Z axis"},FaceGrids→All,ColorFunction→Hue,PlotLabel
→"My 3D Plot",Background→LightYellow,ViewAngle→Pi/7]
Out[40]=
```

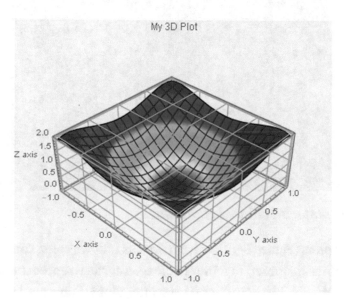

Figure 5-40. *Customized 3D plot*

Hue Color Function and List3D

The Hue color function is a directive that specifies that the values are colored depending on the height they are at. There are three arguments for the Hue color function. First is for the tone of the color (hue); the second marks the saturation; the third marks the bright one; and the fourth the opacity. With hue it is possible to adequately identify the high and low areas from a graph (Figure 5-41) in terms of the four previous features. We can mark these four different parameters. The hue parameters are in the range of 0 to 1.

```
In[41]:= Plot3D[Sin[0.9(x^2+y^3)]/0.5,{x,-1,1},{y,-1,1},FaceGrids→None,
ColorFunction→ (Hue[0.5,1,0.6,0.5]&),PlotLabel→Style["My 3D Plot",Italic,
"Arial"],Background→Black]
Out[41]=
```

195

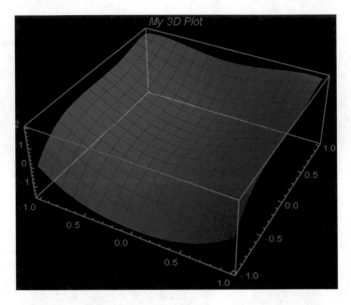

Figure 5-41. *3D plot with colored Hue values*

For 3D scatter plots (Figure 5-42), we can do it as follows using the same data. With ListPlot3D, the points are joined together to create a surface represented by values of the height of each point. With ListPointPlot3D, a scatter plot is generated in 3D points.

```
In[42]:=
 Row[
{
ListPlot3D[Table[RandomReal[1,5],{i,5}],ColorFunction→"SunsetColors",Ticks
→None,PlotLegends→BarLegend[Automatic,LegendMarkerSize→90],ImageSize→
Small,PlotLabel→"ListPlot3D",Filling→Bottom,BoxRatios→Automatic],ListPoi
ntPlot3D[Table[RandomReal[1,5],{i,5}],ColorFunction→"Rainbow",PlotLegends
→BarLegend[Automatic,LegendMarkerSize→90],ImageSize→Small,PlotLabel→"
ListPointPlot3D",Filling→Bottom,BoxStyle→Thick,BoxRatios→{1, 1, 1}]
},Background→Lighter[Gray, 0.80],Frame→True]
Out[42]=
```

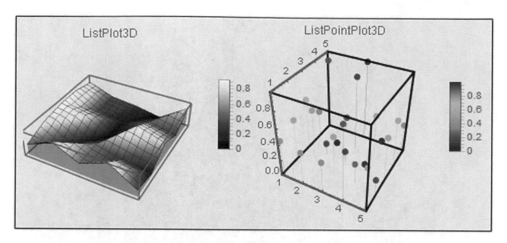

Figure 5-42. *ListPlot3D and ListPointPlot3D for random real numbers*

Contour Plots

One way to visualize a two-variable function is to use a scalar field in which the scalar $z = f(x, y)$ is mapped to the point (x, y). A scalar field can be characterized by its contours (or contour lines) along which the value of $f(x, y)$ is constant. The trace lines of contour line plots or contours can be done using the ContourPlot command, like in the next example.

```
In[43]:= ContourPlot[-((Pi*x)/(3 + x^2 + y^2)),{x,-5,5},{y,-5,5},
ColorFunction→"Temperature",PlotLegends→Automatic,FrameLabel→{x,y}]
Out[43]=
```

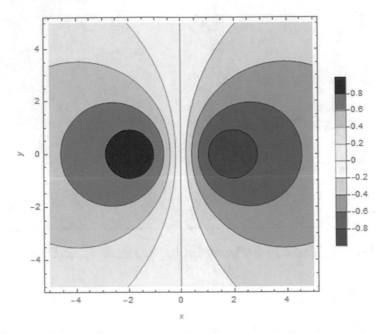

Figure 5-43. *Contour plot for the defined z function*

In Figure 5-43 we plot contour plot, with the options of ColorFunction and
PlotLegends. When we use PlotLegends we specify what type of legends the chart should
use; in this case we use automatic. This shows us the scale of the contours depending
on the color of each outline; for example, for the red is when it is at 0.8 or greater. When
we pass the cursor through the contour curves, the value of that curve will appear. To
label the values of the contour curves in the graph image, add the ContourLabels option
and assign the value to true, as shown in Figure 5-44. To add lines that pass through the
graph, it is possible to use the "GridLines" command, as we saw earlier, or use Mesh.
Mesh can be joined with MeshFunction or MeshStyle.

```
In[44]:= ContourPlot[-((Pi*x)/(3 + x^2 + y^2)),{x,-5,5},{y,-5,5},ColorFunct
ion→"DeepSeaColors",PlotLegends→Automatic,FrameLabel→{x,y},ContourLabels
→True,Mesh→{10,10},MeshStyle→{White}, MeshFunctions→{#3&}]
Out[44]=
```

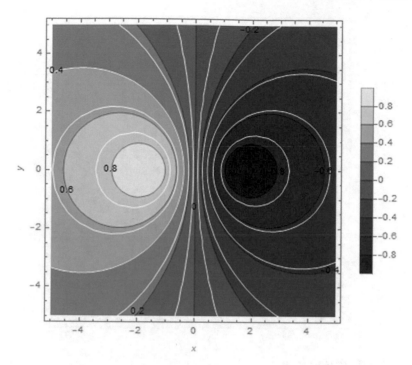

Figure 5-44. *Contour lines added to the contour plot*

To plot data into a contour plot (Figure 5-45), use ListContourPlot. ListContourPlot creates a contour plot out of an array of values shown in heights.

```
In[45]:= ListContourPlot[Table[Exp[x]*Sin[y],{x,0,2,.1},{y,0,2,.1}],Contour
Lines→True,Mesh→Full,ContourLabels→True]
Out[45]=
```

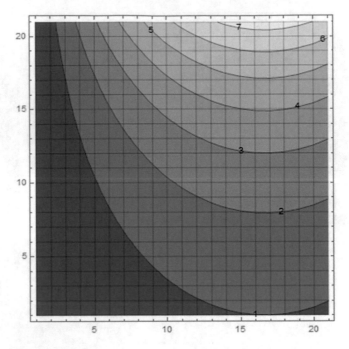

Figure 5-45. *ListContourPlot*

Another helpful plot is DensityPlot (Figure 5-46). DensityPlot works similarly to ContourPlot.

```
In[46]:= DensityPlot[(Sin[2x]*Cos[3y])/5,{x,0,5},{y,0,5},ColorFunction→
"SunsetColors",Mesh→Full]
Out[46]=
```

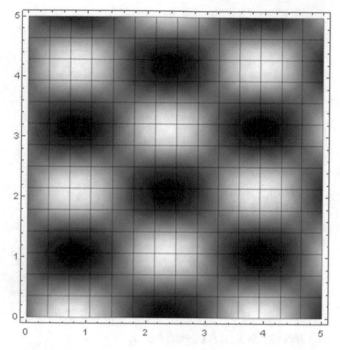

Figure 5-46. *Density plot*

You can plot density plots from data with ListDensityPlot (Figure 5-47).

```
In[47]:= ListDensityPlot[Table[x/3+Sin[3x+y^2],{x,0,5,0.1},{y,0,5,0.1}],
ColorFunction→"LightTemperatureMap",Mesh→10,PlotLegends→Placed[Automatic,
Left]]
Out[47]=
```

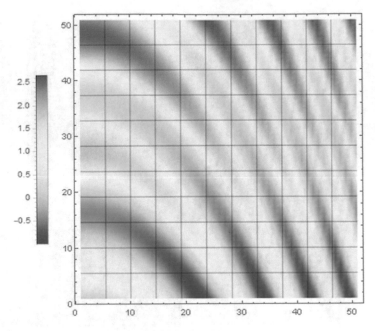

Figure 5-47. *Data represented as a density plot*

3D Plots and 2D Projections

With the Wolfram Language, it is possible to plot functions in 3D and at the same time project the contour maps to planes as the axis, as shown in Figure 5-48.

```
In[48]:= Show[Plot3D[(Sin[2x]*Cos[2y])/4,{x,0,2},{y,0,2},PlotStyle→Directive
[Opacity[1]],AxesLabel→{"X axis","Y axis","Z axis"},ColorFunction→
"Rainbow",PlotTheme→"Marketing"],SliceContourPlot3D[(Sin[2x]*Cos[2y])/4,
{z==-0.15,z==0.15},{x,0,2},{y,0,2},{z,-1,1},ColorFunction→"Rainbow",
Boxed→False],ViewPoint→{1,-1,1}]
Out[48]=
```

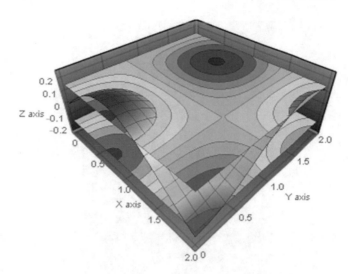

Figure 5-48. *3D plot with contour plots along the xy plane*

Let us point out what happens in the aforementioned code. We plot a function
in 3D (Figure 5-48), and to this function we add color, using the Command Directive
to define the type of opacity, which is set to 1. This is followed by typing the name of
the corresponding axes for the x, y, z axes. The ColorFunction option can help define
a function for the color type; in our case it is Rainbow. The PlotTheme is an option
to plot with various themes for visualization. Coming to this point, we move on to
the SliceContourPlot3D, which gives us a graph of the function, either on a plane or
a surface. If we look, we have plotted when z is worth ± 0.15. A cut is made on the xy
plane. this occurs when x and y are in the range of 0 to 2 and z is in the range of -1 to 1.
In the end we combine the two graphs with the Show command; we use this command
because only by plotting on its own slice contour plot, we would not have the graph of
the function in 3D.

Plot Themes

Preconstructed themes can be accessed with the use of the option "PlotTheme." When
we add the "PlotTheme" option, followed by the first apostrophe, we will see the
autocomplete menu. Figure 5-49 shows the different themes that exist.

Figure 5-49. *PlotTheme pop-up menu*

PlotTheme supports 3D plots, as can be seen in Figure 5-50.

```
In[49]:= Data=Flatten[Table[{x,y,Sin[10(x^2+y^2)]/10},{x,-2,2,0.2},
{y,-2,2,0.2}],1];ListPointPlot3D[Data,ColorFunction→"LightTemperatureMap",
PlotTheme→"Detailed",ViewPoint→{0, -2, 0},ImageSize→250,PlotLegends→
Placed[BarLegend[Automatic,LegendMarkerSize→90],Left],ImageSize→20]
Out[49]=
```

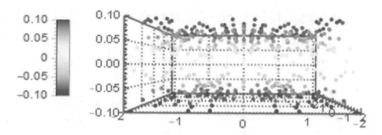

Figure 5-50. *3D scatter plot*

These themes can be used for both third- and second-dimension graphics. Now let us look at another type of theme for a two-dimensional chart (Figure 5-51).

```
In[50]:= Plot[Cos[x],{x,0,10},PlotLabel→"Cos[x]",PlotTheme→"Detailed"]
Out[50]=
```

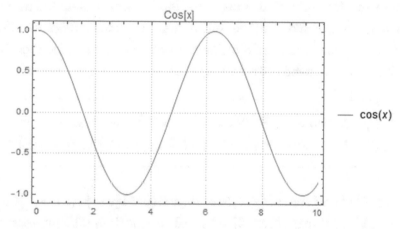

Figure 5-51. *2D plot theme "Detailed"*

I would like to point out a characteristic of PlotTheme. Some themes already have functions within these themes. As we see in the graphic in Figure 5-51, the Detailed theme adds frames, plot legends, and grid lines, even though we can also add them manually.

It is also notable that other topics can only be used for explanatory and demonstrative purposes—that is, no extra information is needed on the chart, but you need to be able to express the information effectively and concretely, as in the Business and Minimal themes (Figure 5-52).

```
In[51]:= Table[Plot[Cos[x],{x,0,10},PlotLabel→"Cos[x]",PlotTheme→Pl],{Pl,
{"Business","Minimal"}}]
Out[51]=
```

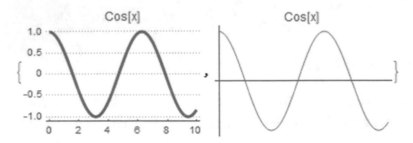

Figure 5-52. *Business and Minimal plot themes*

While there are also topics that show more details like the Detailed theme we saw earlier, other themes exist, like the Scientific theme, as shown in Figure 5-53. At the time, we can add more options, such as ColorFunction and a view with the ViewProjection option, which allows us a fixed observation point.

Note   PlotLegends can work together with ColorFunction, so it shows us how the colors of the dots transition between blue and red, from lowest to highest.

```
In[52]:= Data=Flatten[Table[{x,y,Sin[10(x^2+y^2)]/10},{x,-2,2,0.2},
{y,-2,2,0.2}],1];ListPointPlot3D[Data,ColorFunction→"LightTemperatureMap",
PlotLegends→Placed[BarLegend[Automatic,LegendMarkerSize→90],Left],
PlotTheme→"Scientific",ViewProjection→"Orthographic"]
Out[2]=
```

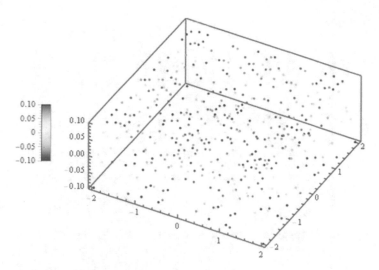

Figure 5-53. *Orthographic point of view*

If what we want is to observe through the coordinate measurements, the option to use is Viewpoint. This option is governed by the following: {x coordinate, y coordinate, z coordinate}. These coordinates are relative to the center of the graph, as shown in the Figure 5-54.

```
In[53]:= ListPointPlot3D[Data,ColorFunction→"LightTemperatureMap
",PlotLegends→Automatic,PlotTheme→"Scientific",ViewPoint→{0,0,-
2},ImageSize→Medium]
Out[53]=
```

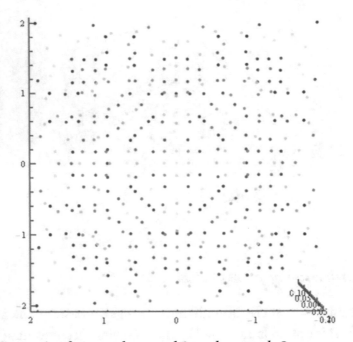

Figure 5-54. *Viewpoint for x and y equal 0 and z equal -2*

CHAPTER 6

Statistical Data Analysis

In the chapter, we will review concepts and techniques to carry out an analysis with the Wolfram Language, as well as perform a linear adjustment through equations and implement specialized functions of the Wolfram Language for the same purpose. With the use of statistical functions. The Wolfram Language is a useful tool for statistics and probability. Mathematica has the functions to perform numerical and approximate calculations for descriptive statistics and random distributions, random numbers and random sampling methods, as we will see on this section.

Random Numbers

In this section we will review the basic commands to generate random numbers—for the case of integers, real and complex. We will see the functions to perform a random sampling with replacement and without replacement in addition to how to make the results reproducible for random numbers.

To create random numbers, there are several functions to generate random integers and real ones. The RandomInteger function will generate entered random numbers; if no arguments are entered in the function, the generation interval is 0 or 1.

```
In[1]:= RandomInteger[]
Out[1]= 0
```

To enter a range, you must define it within the function; for example, between -1 and 1.

```
In[2]:= RandomInteger[{-1,1}]
Out[2]= 1
```

© Jalil Villalobos Alva 2021
J. Villalobos Alva, *Beginning Mathematica and Wolfram for Data Science*,
https://doi.org/10.1007/978-1-4842-6594-9_6

To generate a list of random numbers, we need to define how many numbers within the list we want.

```
In[3]:= RandomInteger[{-1,1},7]
Out[3]= {-1,0,1,-1,-1,-1,1}
```

To repeat the numbers, as a second argument add the form of the list or nested list. For example, create a nested list of seven total items in which each sublist has four items.

```
In[4]:= RandomInteger[{-10,10},{7,4}]
Out[4]= {{-9,-2,6,10},{5,10,3,-7},{7,-2,-4,10},{-1,-2,8,-6},{-10,9,3,0},
{-4,-9,-2,5},{3,1,10,-5}}
```

The function for generating random numbers with a decimal point is called RandomReal. It works similarly to RandomInteger, where the interval is written between curly braces.

```
In[5]:= RandomReal[]
Out[5]= 0.946141
```

There also exists a command for complex random and prime numbers.

```
In[6]:= RandomComplex[]
Out[6]= 0.411636 +0.79253 I
```

For random prime numbers, we must define a minimum and maximum interval— for example, if it is prime number of the first 100.

```
In[7]:= RandomPrime[{1,100},6]
Out[7]= {43,11,61,83,61,79}
```

This type of function generates pseudorandom numbers so that you can set a seed to generate the numbers. This is done with SeedRandom. With a seed we can make sure that the starting sequence of random numbers generated is the same, in order to make random outputs reproducible. To set a seed, use the SeedRandom command. In the following example, we will set a seed followed by a sequence of random numbers; once the seed is introduced, the results should be the same for that seed.

```
In[8]:= SeedRandom[6467789];RandomInteger[{-1,1},3]
Out[8]= {0,1,-1}
```

The seed must go in the same code block to generate the results. There is the option to choose the method, as in the following example where we choose the MersenneTwister method, which is a method commonly used to generate random numbers. Using another method gives us the possibility to generate sequences of different random numbers.

```
In[9]:= SeedRandom[Method→"MersenneTwister"];RandomInteger[{-1,1},{3,3}]
//MatrixForm
Out[9]//MatrixForm=
```

$$\begin{pmatrix} 1 & -1 & 1 \\ 1 & 0 & 0 \\ 0 & -1 & -1 \end{pmatrix}$$

To return to the original value, the seed enters the function without arguments.

```
In[10]:= SeedRandom[]
```

In addition to introducing a seed, we can create blocks of random numbers in which functions can be used locally and not affect random behavior outside these blocks. This is done with the BlockRandom function.

```
In[11]:= BlockRandom[RandomReal[1]]
Out[11]= 0.943218
```

If we run an algorithm that produces random numbers within the BlockRandom and declare our own seed, this should not impact other processes where random numbers are generated outside the BlockRandom. To illustrate, let us look at the example.

```
In[12]:=
SeedRandom[121];
{RandomReal[],BlockRandom[RandomReal[]],RandomReal[],RandomReal[]}
Out[12]= {0.0908251,0.194288,0.194288,0.296762}
```

As seen, the latter process generated different random numbers

Random Sampling

To make a sample with a replacement, the function to use is RandomChoice. To select a single item, we write only the list. We will set a seed to get the same results.

```
In[13]:= SeedRandom[12345];
RanData=RandomReal[{0,1},10]
Out[13]= {0.121246,0.329922,0.782753,0.430168,0.223586,0.463053,0.738017,
0.707618,0.790911,0.105714}
```

We have generated a list of 10 random numbers in the range from 0 to 1, and now we proceed to randomly choose an item of these numbers.

```
In[14]:= RandomChoice[RanData]
Out[14]= 0.738017
```

This gives us a single result from the list of 10 items. Similarly, we can choose the number of samples with a number of elements, with the following form: RandomChoice["data", "number of samples", "number of elements"]. Of the 10 elements, we will now pick three samples with one element.

```
In[15]:= RandomChoice[RanData,{3,1}]
Out[15]= {{0.329922},{0.738017},{0.223586}}
```

Although, if we want it in the same sample, we only need to specify the number of elements to choose.

```
In[16]:= RandomChoice[RanData,5]
Out[16]= {0.790911,0.463053,0.329922,0.430168,0.329922}
```

To get a sampling without replacement use RandomSample. This function does not choose a list item from the data list more than once. To choose we only specify the number of elements in the sample as the second argument, since the first one corresponds to the data list.

```
In[17]:= RandomSample[RanData,9]
Out[17]= {0.105714,0.790911,0.463053,0.738017,0.707618,0.782753,0.430168,0.
223586,0.329922}
```

Looking in detail, we notice that there is no repeated value. Each item in the list has an equal probability of being selected in sampling.

In the case that each item in the list has a specific weight associated with it, then to enter those terms we use the following form of expression, {w1, w2, w3...} → { element1, element2, element3...}; the list of items is associated with a specific weight for replacement sampling.

We will denote the list of weights and do the sampling by associating the weights and elements.

```
In[18]:= W={0.03`,0.08`,0.22`,0.04`,0.12`,0.3`,0.12`,0.03`,0.04`,0.02`};
RandomChoice[W→RanData,2]
Out[18]= {0.223586,0.430168}
```

As we notice, they are chosen depending on how each element is assigned a weight. For sampling without replacement, the process is analogous.

```
In[19]:= RandomSample[W→RanData,3]
Out[19]= {0.463053,0.738017,0.223586}
```

Systematic Sampling

To perform a system sampling we must determine the size of the sample, M. To get the sample size, we can list the items in the list or get the length of the list. To get started, we will create a list of 200 prime numbers.

```
In[20]:= SeedRandom[09876]
RPrime=RandomPrime[{1,100},200];
Length[RPrime]
Out[20]= 200
```

We have already calculated the sample size, so we must determine the size of a specific sample; for this case we want a sample of 20 elements. Once the sample is determined, we will calculate the interval of the denoted sampling j; j is calculated through a ratio, the original sample size divided by the total number of elements in the specified sample.

```
In[21]:= j=Length[RPrime]/20
Out[21]= 10
```

This means that the sampling interval for our new sample will be from 1 to 10. From here, we select a random number within the interval, and from there we add j times to choose the next element; that is, for the first element it will be a random h number of the range [1,10], for the second it will be h + j, and for the third h +3j, and so on, until it reaches the size of the original sample.

We chose a random number between 1 and 10.

```
In[22]:= RandomSample[Range[10],1]
Out[22]= {9}
```

The result means that we select from the ninth element. We deploy the list to have a better view of the data.

```
In[23]:= RPrime
Out[23]= {17,11,67,97,11,73,71,61,71,31,59,29,79,7,71,89,79,11,2,29,97,61,
2,71,3,79,31,83,83,17,37,89,41,31,61,7,11,53,17,61,71,2,53,23,29,59,11,41,
13,71,3,53,13,61,19,2,17,17,59,3,11,41,83,59,41,47,13,59,17,5,5,59,79,37,
97,7,11,23,41,83,67,79,73,73,73,41,79,17,59,37,83,71,73,17,2,11,41,89,97,
7,2,23,13,67,79,83,5,61,47,73,61,97,53,53,2,89,19,19,61,89,83,43,73,3,83,
17,5,89,29,23,7,23,53,97,2,83,13,17,37,2,19,59,79,29,43,19,7,43,59,47,3,41,
23,53,37,59,29,83,37,59,19,59,31,89,2,67,47,47,97,2,47,97,41,11,43,37,7,59,
67,83,89,2,17,13,2,7,73,83,89,2,3,59,17,19,73,13,53,29,89,83}
```

To get the positions of the items to be selected, it would be the random number for the selection, which is 9, plus n times j until you have 20 elements.

```
In[24]:= Table[9+n*j,{n,0,19}]
Out[24]= {9,19,29,39,49,59,69,79,89,99,109,119,129,139,149,159,169,179,189,
199}
```

Note   Remember that position index starts from 1 to n elements.

We must choose the positions shown in the previous output. To choose, we will use the double square bracket notation.

```
In[25]:= Table[RPrime[[9+n*j]],{n,0,19}]
Out[25]= {71,2,83,17,13,59,17,41,59,97,47,61,29,37,59,37,97,67,89,89}
```

Let us take a better look at the selected elements, highlighting them in red with the help of MapAt and Style.

```
In[26]:= MapAt[Style[#,FontColor→ColorData["HTML"]["Red"]]&,RPrime,{#}&/@
{9,19,29,39,49,59,69,79,89,99,109,119,129,139,149,159,169,179,189,199}
Out[26]= {17,11,67,97,11,73,71,61,71,31,59,29,79,7,71,89,79,11,2,29,97,61,
2,71,3,79,31,83,83,17,37,89,41,31,61,7,11,53,17,61,71,2,53,23,29,59,11,41,
13,71,3,53,13,61,19,2,17,17,59,3,11,41,83,59,41,47,13,59,17,5,5,59,79,37,
97,7,11,23,41,83,67,79,73,73,73,41,79,17,59,37,83,71,73,17,2,11,41,89,97,7,
2,23,13,67,79,83,5,61,47,73,61,97,53,53,2,89,19,19,61,89,83,43,73,3,83,17,
5,89,29,23,7,23,53,97,2,83,13,17,37,2,19,59,79,29,43,19,7,43,59,47,3,41,23,
53,37,59,29,83,37,59,19,59,31,89,2,67,47,47,97,2,47,97,41,11,43,37,7,59,67,
83,89,2,17,13,2,7,73,83,89,2,3,59,17,19,73,13,53,29,89,83}
```

As we can see, system sampling does not create a completely random sample. The random selection comes in the first part when we select the first item to create the new sample. Once the first item is selected, the other selections are from a succession of non-random numbers. Another aspect to consider is the order of the original sample; if there is periodicity between the elements this can lead to a great variability in the selecting of elements.

Common Statistical Measures

Grasping the commonly used statistical formulas are crucial to understanding how the data is behaving on set of conditions. Descriptive statistics are implemented once data has been collected and is one of the first steps in the process of exploratory data analysis, which allows you to find insights of the data collected in terms of discovering patterns, anomalies, trends, seasonality, variations, etc.

Exploratory data analysis is depicted as a set of analysis techniques with the purpose of detecting characteristics that are not visible at first sight or revealed once the data has been collected. The basic structure of this technique relies on numeric data analysis, graphical representation, and a statistical model. Many reasons to use data exploratory analysis include reviewing for missing data, describing a general and particular idea of the underlying structure, and analyzing for different assumptions associated with the model creation, among many more.

The proposal of such a process was introduced by Jhon Tukey in 1977. To review more depth about this technique, visit the following reference, *Exploratory Data Analysis* (Tukey, J. W. [1977], Vol. 2, pp. 131-160).

Measures of Central Tendency

Given a sample of data, we can calculate the descriptive measures. Central trend measures are those parameters that give us information on the average values of the data to be studied.

The mean, also known as arithmetic mean, is a parameter that is calculated from the sum of the values of the sample and dividing by the sum of the number of elements. The Mean function calculates the average.

```
In[27]:= List1=Table[Prime[i],{i,10}];
"Prime list :" <>ToString@ List1
"Mean: "<>ToString@Mean@N@List1
Out[27]= Prime list :{2, 3, 5, 7, 11, 13, 17, 19, 23, 29}
Out[27]= Mean: 12.9
```

Note   The symbol <> is the short notation for StringJoin.

The median is the value that divides the sample into two equal parts, since it is the midpoint of the data, so the median is the symmetry value relative to the number of data. The Median function gives us this value.

```
In[28]:= "Median: "<>ToString@Median@List1
Out[28]= Median: 12
```

Mode is the most common value of the sample. We use the Counts command, which gives us the number of occurrences of each item in the list.

```
In[29]:= Counts[List1]
Out[29]= <|2→1,3→1,5→1,7→1,11→1,13→1,17→1,19→1,23→1,29→1|>
```

In this case, the occurrence is 1. There are no repeated values; we can say that there is no mode in this data sample.

Measures of Dispersion

Dispersion measurements reveal information on the variability presented in the sample. The range tells us the interval in which the data varies. This is taken by subtracting the max value and the minimum value. The Max and Min functions return the maximum and minimum value of a list.

```
In[30]:= "Range: "<>ToString[Max[List1]-Min[List1]]
Out[30]= Range: 27
```

Variance is a measure obtained from the subtraction of the mean to each of the elements of the sample. The result is squared followed by adding them together. The summation is divided by the size of the sample. Its function is Variance.

```
In[31]:= "Variance: "<>ToString[N[Variance[List1],3]]
Out[31]= Variance: 81.4
```

Standard deviation is a measurement obtained from the square root of the variance or by means of the StandardDeviation function.

```
In[32]:= {"Square root of Variance: " <>ToString[N[Sqrt[Variance[List1]],2]],
"StandardDeviation: " <>ToString[N[StandardDeviation[List1],2]]}
Out[32]= {Square root of Variance: 9.0,StandardDeviation: 9.0}
```

Standard score is a score called z and measures how many standard deviations are away from the arithmetic average for each element of the sample. The mathematical equation is $z = \dfrac{x - \mu}{\sigma}$, where x is the measure, μ the mean, and σ the standard deviation. If z is positive, it means that that element is greater than the mean. When z is negative, it is the opposite case. We will determine the z-score for the second item in the list.

```
In[33]:= z=N[(List1[[2]]-Mean@List1)/StandardDeviation@List1,3];
"z score: " <> ToString@z
Out[33]= z score: -1.10
```

This means that the score for the second element is 1.10 times below average.

Quartile calculation divides data into four equal parts. The lower quartile corresponds to the 25% quartile of the data, while the second quartile is 50%, the third quartile (the upper quartile) is 75%, and the fourth quartile (100%). To calculate the quartiles, we use the Quartiles function, which in turn gives the values of the first, second, and third quartile.

```
In[34]:= "Quartiles: " <> ToString@Quartiles[List1]
Out[34]= Quartiles: {5, 12, 19}
```

If we want to get the single value, we use the Quantile function, followed by the percentile to be calculated. Then for the calculation of the third quartile (75th percentile) we use the following for.

```
In[35]:= Quantile[List1,0.75]
Out[35]= 19
```

To calculate the interquartile range, which is the difference between the upper and lower quartiles, the function is InterquartileRange.

```
In[36]:= InterquartileRange[List1]
Out[36]= 14
```

Statistical Charts
BarCharts

Sometimes when we carry out a statistical study it is possible to find quantitative and qualitative variables; for these variables we can create a bar graph representation. A bar graph (Figure 6-1) is a graphical representation where the number of frequencies of a discrete qualitative variable is displayed on an axis.

```
In[37]:= BarChart[{1,2,3,4},ChartLabels→{"feature 1","feature 2",
"feature 3","feature 4"}]
Out[37]=
```

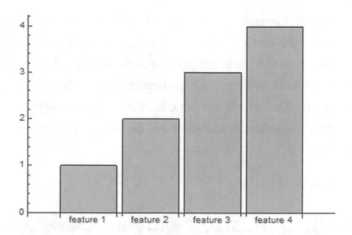

Figure 6-1. *Bar chart*

The different modalities of the qualitative variable are positioned on one of the axes. The other axis shows the value or frequency of each category on a given scale. The feature 2 bar has an associated value of 2. The orientation of the graph can be vertical, where the categories are located on the horizontal axis and the bars are vertical, or horizontal, where the categories are located on the vertical axis and the bars are horizontal (Figure 6-2).

```
In[38]:= GraphicsRow[{BarChart[{1,2,3,4},ChartLabels→{"feature 1",
"feature 2","feature 3","feature 4"},BarOrigin→Bottom,ChartStyle→
LightBlue],BarChart[{1,2,3,4},ChartLabels→{"feature 1","feature 2",
"feature 3","feature 4"},BarOrigin→Left,ChartStyle→LightRed]}]
Out[38]=
```

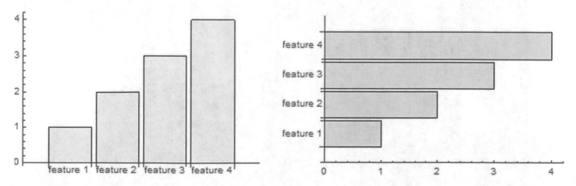

Figure 6-2. *Bottom and left origin bar chart*

Bar graphs can be used to compare magnitudes of different categories and observe how values change according to a fixed variable—for example, each feature. In addition, we can choose how to show the bars: simply, where we show a single series, shown in the earlier example; grouped, which contains several data series and is represented by a different type of bar; or stacked, where the bar is divided into segments with different colors representing various categories. Percentile layout is displayed on a percentage scale, as shown in Figure 6-3.

```
In[39]:= Labeled[GraphicsGrid[
{{
BarChart[{{4,3,2,1},{1,2,3},{3,5}},ChartLayout→"Grouped",ColorFunction→
"SolarColors"],
BarChart[{1,2,3,4},ChartStyle→LightRed,ChartLayout→"Stepped"]},
{BarChart[{{4,3,2,1},{1,2,3},{6,5}},ChartLayout→"Stacked"],
BarChart[{{4,3,2,1},{1,2,3},{6,5}},ChartLayout→"Percentile",ColorFunction→
"DarkRainbow"]
}},Frame→All,FrameStyle→Directive[Black,Dashed],Background→LightBlue,
ImageSize→500],"Bar Charts",Top]
Out[39]=
```

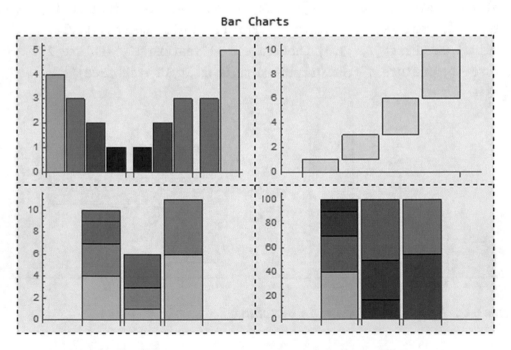

Figure 6-3. *Bar chart grid*

There is also the counterpart to 3D graphics, with BarChart3D (Figure 6-4).

```
In[40]:= SeedRandom[123]
Labeled[GraphicsGrid[
{{
BarChart3D[{{4,3,2,1},{1,2,3},{3,5}},ChartLayout→"Grouped",ColorFunction→
"SolarColors"],
BarChart3D[{1,2,3,4},ChartStyle→LightRed,ChartLayout→"Stepped"]},
{BarChart3D[RandomReal[1,{10,5}],ChartLayout→"Stacked"],
BarChart3D[{{4,3,2,1},{1,2,3},{6,5}},ChartLayout→"Percentile",
ColorFunction→"DarkRainbow"]
}},Frame→All,FrameStyle→Directive[Red,Thick],Background→LightBlue,
ImageSize→500],"3D Bar Charts",Top,Frame→True,Background→White]
Out[40]=
```

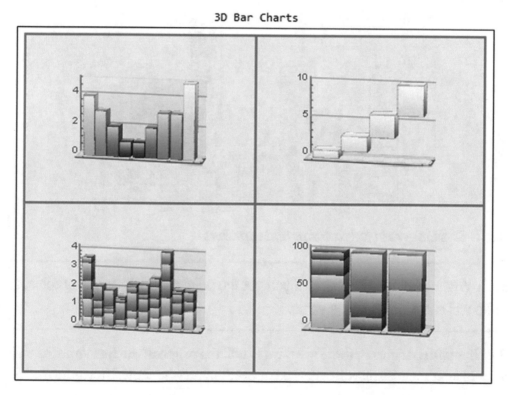

Figure 6-4. *3D bar charts grid*

Histograms

Histograms are a type of visualization that is commonly used in statistical studies. With histograms we can see how a sample is distributed. Histograms are used to represent the frequencies of a quantitative variable. The classes of the variable are positioned on the horizontal axis and the frequencies on the other axis. In the next examples, we will graph a histogram from a population of 50 random values between 0 and 1 and set the number of bins to 10. The second argument to histograms is to define the number of bins (Figure 6-5).

```
In[41]:= SeedRandom[4322]
hist1=Table[RandomReal[{2,3}],{i,0,20}];
Histogram[hist1,10]
Out[41]=
```

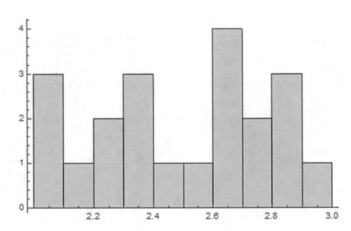

Figure 6-5. *Histogram for random real numbers*

Note   When dealing with charts, if you put the pointer cursor on the graphic, an info tip will show marking the value.

Just like with bar charts, there are ways to edit the origin of the histogram as well as how the histogram is displayed—stacked or overlapped—as shown in Figure 6-6.

```
In[42]:=
hist2=Table[Cos[i],{i,1,20}];
hist3=Table[Sin[i],{i,1,10}];
```

```
GraphicsColumn[{Histogram[{hist1,hist2},10,BarOrigin→Left,ChartStyle→
"Pastel",ChartLegends→{"rand num","Cos(x)"}],Histogram[{hist2,hist3},10,
ChartLayout→"Overlapped",ChartStyle→"Pastel",ChartLegends→{"Cos(x)",
"Sin(x)"}],Histogram[{hist2,hist3},10,ChartLayout→"Stacked",ChartStyle→
"Pastel",ChartLegends→{"Cos(x)","Sin(x)"}]}]
Out[42]=
```

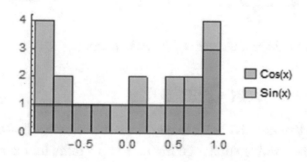

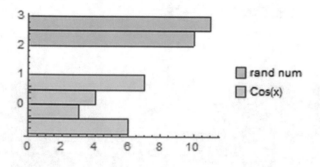

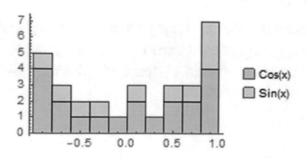

Figure 6-6. *Histogram shapes grid*

With this in mind we can also graph bidirectional histograms using PairedHistograms. These can be horizontal or vertical orientation and contain two data series whose bars go in opposite directions (Figure 6-7).

```
In[43]:= SeedRandom[123]
GraphicsRow[{PairedHistogram[{RandomReal[{0,1},20]},{RandomReal[{0,1},20]},
BarOrigin→Left],PairedHistogram[{RandomReal[{0,1},20]},{RandomReal[{0,1},
20]},10,BarOrigin→Top,ChartStyle→"Pastel"]}]
Out[43]=
```

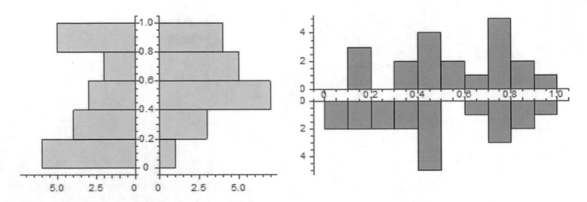

Figure 6-7. *Paired histograms with different origins*

Pie Charts and Sector Charts

Pie charts are circles that are divided into two or more sections. They are used to represent quantitative variables that together make up a total; for example, the size of the sector is drawn proportional to the value it represents and is expressed in percentages, which only provides relative quantitative information. Pie charts are made with the command PieChart (Figure 6-8).

```
In[183]:= GraphicsRow[{PieChart[{1,1,1},ChartLegends→{"part a","part b",
"part c"},ChartStyle→{LightRed, LightBlue, LightYellow}],
PieChart[{1,1},ChartLegends→{"part a","part b"},ChartStyle→"SunsetColors"]}]
Out[183]=
```

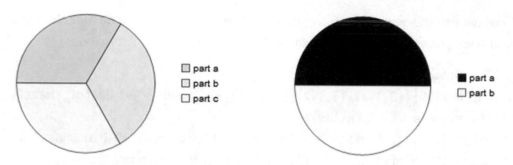

Figure 6-8. *Pie charts*

Sector charts are graphed with the SectorChart command (Figure 6-9). They are used to compare different data that occur in the same place. They are constructed from the proportional size of x to the value of the radius of y. The dimension in which the quantities are expressed must be the same for all the segments.

```
In[45]:= SectorChart[{{2,1},{1,2}},ChartLegends→{"Sector a","Sector b"},
ChartStyle→{LightRed, LightYellow}]
Out[45]=
```

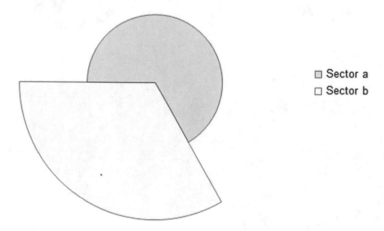

Figure 6-9. *Sector chart*

For each graph seen, there is a corresponding command to create them in three dimensions; these are shown in Figure 6-10.

```
In[46]:= GraphicsGrid[
{{SectorChart3D[{{2,1,1},{3,1,2},{1,2,2}},PlotLabel→"3D Sector chart",
ChartStyle→{Red, Blue,Yellow}],
PieChart3D[{1,1,1},ChartStyle→"GrayTones",PlotLabel→"3D Pie Chart"]},
{Histogram3D[Table[{i^3,i^-1},{i,20}],10,ChartElementFunction→
"GradientScaleCube",PlotLabel→"3D Histogram"],None}
},ImageSize→500,Frame→True,FrameStyle→Directive[Thick,Dotted]]
Out[46]=
```

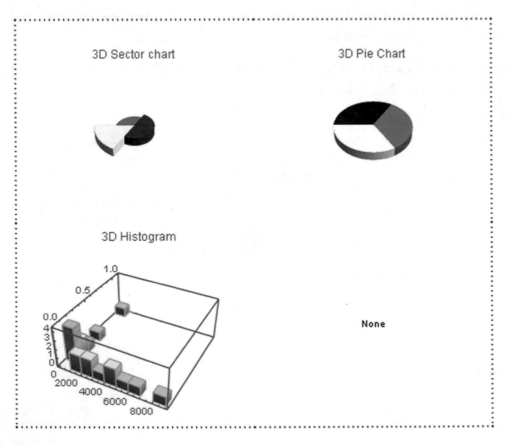

Figure 6-10. *3D grid charts*

Box Plots

The box plot is a way of representing and observing a distribution of data. Fundamentally, it is used to highlight aspects of the distribution of data in one or more series. To graph a box plot, we use the BoxWhiskerChart command (Figure 6-11).

```
In[47]:= SeedRandom[1234]
BoxWhiskerChart[{Table[RandomReal[],{i,0,50}],Table[RandomReal[],{i,0,50}],
Table[RandomReal[],{i,0,15}]},ChartLabels→{"Chart 1","Chart 2","Chart 3"}]
Out[47]=
```

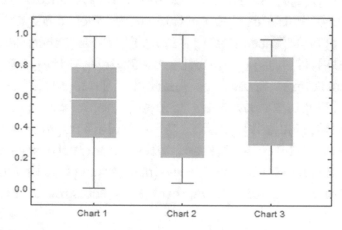

Figure 6-11. *Box plot*

The box is represented by a rectangle that marks the interquartile range of the distribution. The first line from bottom to top marks the value of the first quartile (25%), the line that crosses the box is the median, and the last line that delimits the box is the third quartile (75%). Whiskers are the lines that mark the maximum and minimum values. When passing the mouse cursor over the plot, information about the data will be shown; this includes minimum, maximum, median, 75th percentile, and 1st quartile. Depending on the specification we use, this can affect what parameters are displayed and how (Figure 6-12).

```
In[48]:= SeedRandom[123]
GraphicsGrid[{{BoxWhiskerChart[{Table[RandomReal[],{i,0,50}],
Table[RandomReal[],{i,0,50}],Table[RandomReal[],{i,0,15}]},#,
PlotLabel→Style[#,White],
```

```
ImageSize→Medium,ChartStyle→"MintColors",FrameStyle→Directive[White,12]]
&["Median"],
BoxWhiskerChart[{Table[RandomReal[],{i,0,50}],Table[RandomReal[],{i,0,50}],
Table[RandomReal[],{i,0,15}]},#,PlotLabel→Style[#,LightOrange],ImageSize→
Medium,ChartStyle→"MintColors",FrameStyle→Directive[Orange,12]]&["Basic"],
BoxWhiskerChart[{Table[RandomReal[],{i,0,50}],Table[RandomReal[],{i,0,50}],
Table[RandomReal[],{i,0,15}]},#,PlotLabel→Style[#,White],ImageSize→Medium,
ChartStyle→"MintColors",FrameStyle→Directive[White,12]]}&["Notched"],
{BoxWhiskerChart[{Table[RandomReal[],{i,0,50}],Table[RandomReal[],{i,0,50}],
Table[RandomReal[],{i,0,15}]},#,PlotLabel→Style[#,LightOrange],ImageSize→
Medium,ChartStyle→"MintColors",FrameStyle→Directive[Orange,12]]&["Outliers"],
BoxWhiskerChart[{Table[RandomReal[],{i,0,50}],Table[RandomReal[],{i,0,50}],
Table[RandomReal[],{i,0,15}]},#,PlotLabel→Style[#,White],ImageSize→Medium,
ChartStyle→"MintColors",FrameStyle→Directive[White,12]]&["Mean"],
BoxWhiskerChart[{Table[RandomReal[],{i,0,50}],Table[RandomReal[],{i,0,50}],
Table[RandomReal[],{i,0,15}]},#,PlotLabel→Style[#,LightOrange],ImageSize→
Medium,ChartStyle→"MintColors",FrameStyle→Directive[Orange,12]]&[
"Diamond"]}},FrameTicksStyle→18,Frame→{None,None,{{1,1}→True,{2,2}→True,
{1,3}→True}},FrameStyle→Directive[Thick,Red],Background→Black]
Out[48]=
```

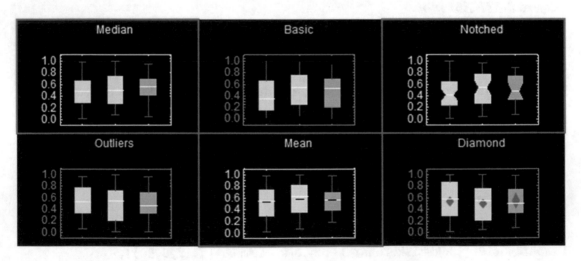

Figure 6-12. *Multiple box plots*

Median is the default specification; it shows the median in the center of the box. Basic is to show only the box. Notches show the confidence interval for the median. Outliers shows and marks the atypical points. Mean marks the mean of the distribution, and Diamond notes the confidence interval for the mean.

Distribution Chart

A violin diagram is used to visualize the distribution of the data and the probability density. To plot a violin plot (Figure 6-13), the DistributionChart command is used.

```
In[49]:= DistributionChart[Table[i^Exp[i],{i,0,1,0.01}]]
Out[49]=
```

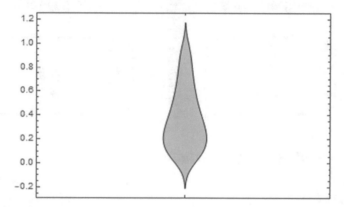

Figure 6-13. *Violin plot*

The graph, shown in the figure, is a combination of a box-and-whisker plot and a density plot on each side to show how the data is distributed. DistributionChart has different shapes to graph (Figure 6-14).

```
In[50]:= GraphicsGrid[{{DistributionChart[Table[i^Exp[i],{i,0,2,0.1}],
ChartElementFunction→"SmoothDensity",PlotLabel→"SmoothDensity"],
DistributionChart[Table[i^Exp[i],{i,1,2,0.1}],ChartElementFunction→
"Density",PlotLabel→"Density",FrameStyle→Directive[Red,12]]},{
DistributionChart[Table[i^Exp[i],{i,0,1,0.09}],ChartElementFunction→
"HistogramDensity",PlotLabel→"HistogramDensity",FrameStyle→Directive[Red,
12]],DistributionChart[Table[i^Exp[i],{i,0,1,0.0112}],ChartElementFunction→
```

```
"PointDensity",PlotLabel→"PointDensity"]}},ImageSize→Medium,FrameStyle→
Directive[Thickness[0.02],LightGray],Dividers→{2→Directive[Black,Dotted],
2→Directive[Black,Dotted]},Frame→{1→False,False}]
Out[50]=
```

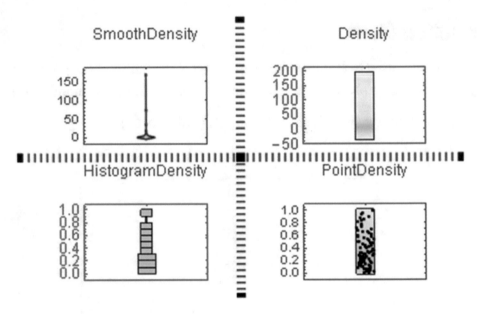

Figure 6-14. *Violin plots in different shapes*

Charts Palette

Another way to add options to charts is through the Chart Element Schemes palette, which is found within the Palettes menu (Palettes → Chart Element Schemes). This palette is shown in Figure 6-15.

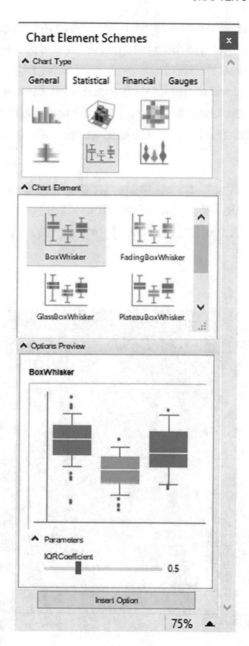

Figure 6-15. *Chart Element Schemes palette*

In the palette, you will find three categories. Chart Type is where we choose the type of chart. This contains four tabs: (1) general, where the graphics are found from bar charts, sector, footer, and others; (2) statistical graphs associated with data distributions; (3) financial, associated with charts for financial data; and (4) gauges, which are diagrams of measures. The second category is to choose the shape of the graph with the ChartElemenFunction option. The third category is for the preview of the options chosen from the previous categories.

To illustrate this, let us look at the following exercises. First, we will make the graph of the density of a histogram, and later we will modify the shape of the graph with the help of the palette. To graph the density of a histogram, we use the DensityHistogram command (Figure 6-16).

```
In[51]:= DensityHistogram[Flatten[Table[{x^2+y^2,x^2-y^2},{x,0,2,0.1},{y,0,
2,0.1}],1],ChartBaseStyle→Red,ColorFunction→"SolarColors",Background→
Black,FrameStyle→Directive[White,Thick],FrameLabel→{"X","Y"},ImageSize→
300]
Out[51]=
```

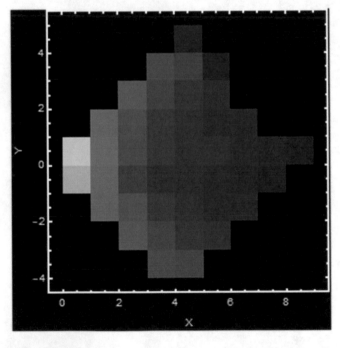

Figure 6-16. *Density histogram*

Once the graph is done, we will add an option with the pallet head and open the Chart Element Schemes palette. Within chart type, we click the statistical tab, and we will choose the DensityHistogram chart. Once the chart has been selected, we now go to Chart Element and select that the type of form is Bubble. Then we go to Options Preview to see how our graph would look; if we click Shape, a pop-up menu will appear with other shapes; we choose hexagon. Figure 6-17 shows how the preview of the selected chart elements should look.

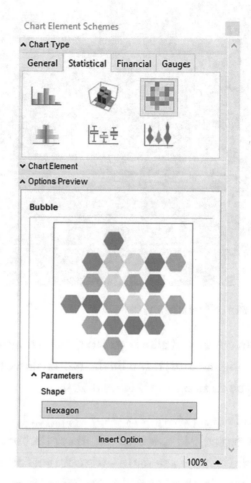

Figure 6-17. *Density histogram options selected*

Once finished selecting, we click the insert button so that it inserts the following code: ChartElementFunction → ChartElementDataFunction ["Bubble", "Shape" → "Hexagon"]. To graph it properly, we add this code as an option and proceed to plot it (Figure 6-18) to observe the new option added.

```
In[52]:= DensityHistogram[Flatten[Table[{x^2+y^2,x^2-y^2},{x,0,2,0.1},{y,0,
2,0.1}],1],ChartBaseStyle→Red,ColorFunction→"SolarColors",Background→
Black,FrameStyle→Directive[White,Thick],FrameLabel→{"X","Y"},ImageSize→300,
ChartElementFunction→ChartElementDataFunction["Bubble","Shape"→"Hexagon"]]
Out[52]=
```

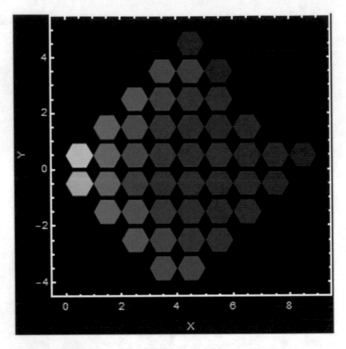

Figure 6-18. *Hexagon density histogram*

The DensityHistogram command allows you to choose how to display the distribution of the data along the axes; it can be the dimensions, box plots, or histograms if we select the Method type as an option (Figure 6-19).

```
In[53]:= Hist=Flatten[Table[{x^2+y^2,x^2-y^2},{x,0,2,0.1},{y,0,2,0.1}],1];
{MenuView[{DensityHistogram[Hist,Method→{"DistributionAxes"→True},
ColorFunction→GrayLevel,ChartBaseStyle→Directive[FaceForm[Opacity[0.5]],
EdgeForm[Red]],ChartLegends→BarLegend[Automatic,LegendMarkerSize→70],
PlotLabel→Style[" Density Histogram 1", Bold],ChartElementFunction→#,
ImageSize→200],
DensityHistogram[Hist,Method→{"DistributionAxes"→"Histogram"},ChartLegends→
BarLegend[Automatic,LegendMarkerSize→70],PlotLabel→Style[" Density
Histogram 2", Bold],ChartBaseStyle→EdgeForm[Thick],PlotTheme→"Scientific",
ChartElementFunction→#,ImageSize→200],
DensityHistogram[Hist,Method→{"DistributionAxes"→"BoxWhisker"},
ColorFunction→"BlueGreenYellow",PlotLabel→Style[" Density Histogram 3",
Bold],ChartLegends→BarLegend[Automatic,LegendMarkerSize→70],
ChartElementFunction→#,ImageSize→200]
}]&[ChartElementDataFunction["Bubble","Shape"→"Hexagon"]],{
```

```
GraphicsRow[{
DensityHistogram[Hist,Method→{"DistributionAxes"→True},ColorFunction→
GrayLevel,ChartBaseStyle→Directive[FaceForm[Opacity[0.5]],EdgeForm[Red]],
ChartLegends→BarLegend[Automatic,LegendMarkerSize→70],PlotLabel→Style["
Density Histogram 1", Bold],ChartElementFunction→#,ImageSize→130],
DensityHistogram[Hist,Method→{"DistributionAxes"→"Histogram"},ChartLegends→
BarLegend[Automatic,LegendMarkerSize→70],PlotLabel→Style[" Density
Histogram 2", Bold],ChartBaseStyle→EdgeForm[Thick],PlotTheme→"Scientific",
ChartElementFunction→#,ImageSize→130],
DensityHistogram[Hist,Method→{"DistributionAxes"→"BoxWhisker"},
ColorFunction→"BlueGreenYellow",PlotLabel→Style[" Density Histogram 3",
Bold],ChartLegends→BarLegend[Automatic,LegendMarkerSize→70],
ChartElementFunction→#,ImageSize→130]
}&[ChartElementDataFunction["Bubble","Shape"→"Hexagon"]]]}}
Out[53]=
```

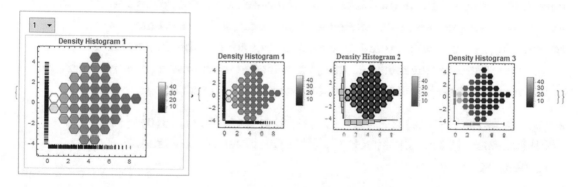

Figure 6-19. *Menu view of the three different method plots*

The plots are shown inside as a menu, so to access the different graphs, you have
to select each graph within the menu. Even so, we show the plots on a small scale to
demonstrate how they should look (Figure 6-19). The first graph shows the dimensions
of the data distribution along the axes. The second shows the distribution of the data in
the form of histograms, and the third shows the box plots.

Ordinary Least Square Method

The method of ordinary least squares basically consists of finding a line that best fits the data. This method is used to study the relationship between the dependent variable and the independent variable. The method is based on the expression of finding a line of the form y = mx + b, where x is the independent variable, y is the dependent variable, m is the slope, and b the y-intercept. The calculation of the slope and the sorted to origin b is obtained from the following equations.

$$m = \frac{n * \sum(x*y) - \sum x * \sum y}{n * \sum x^2 - |\sum x|^2}$$

$$b = \frac{\sum y * \sum x^2 - \sum x * \sum(x*y)}{n * \sum x^2 - |\sum x|^2}$$

The summation is denoted by the Greek capital letter sigma ($\sum$); n is the amount of data in the sample. The method is calculated for measured data pairs and slope values, and y-intercept sources are calculated to create the best fit to the data to a line. By substituting in the general equation, we get the equation of the line for the dataset.

To illustrate what the method is like, let us look at the following example using our points for the dependent variable and the independent variable.

```
In[54]:= Data={{-1,10},{0,9},{1,7},{2,5},{3,4},{4,3},{5,0},{7,-1}};
Grid[Transpose[Prepend[Data,{"X","Y"}]],Dividers→{2→True,2→True},
Alignment→Center]
Out[54]=
```

X	-1	0	1	2	3	4	5	7
Y	10	9	7	5	4	3	0	-1

Now we have to calculate the data needed to get the slope and y-intercept.

```
In[55]:=
n=Length[Data];
SumX=Total@Data[[All,1]];
SumY=Total@Data[[All,2]];
SumXY=Total[Data[[All,1]]*Data[[All,2]]];
SumXSqre=Total@(Data[[All,1]]^2);
```

$$m = N @ \frac{n * \text{SumXY} - \text{SumX} * \text{SumY}}{n * \text{SumXSqre} - \text{Abs}[\text{SumX}]^2};$$

$$b = N @ \frac{\text{SumY} * \text{SumXSqre} - \text{SumX} * \text{SumXY}}{n * \text{SumXSqre} - \text{Abs}[\text{SumX}]^2};$$

To solve the equation of the shape y = mx + b, we use the Solve command. The first argument is the equation, and the second argument is for the variable to solve. To enter the equation, we must use the same double notation, since a single equal is for set instruction.

```
In[56]:= Solve[SetPrecision[y==m*x+b,3],y]
Out[56]= {{y→8.47-1.47 x}}
```

This results in the equation of the line being y = 1.47 x + 8.47. Given this equation, we will plot the points and the line that best fits these points (Figure 6-20).

```
In[57]:=
Show[Plot[b+m x,{x,-1,8},PlotLegends→Placed[" Linear Fit: y=-1.47 x +
8.47",{0.6,0.8}],PlotRange→Automatic],ListPlot[Data,PlotStyle→Red]]
Out[57]=
```

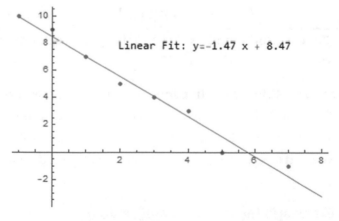

Figure 6-20. *Plot of data and fitted curve*

Having obtained the equation, we observe that this is a model with negative slope; this is corroborated by the graph of the equation shown in blue.

Pearson Coefficient

The measure that tells us that both the points fit to the equation is the Pearson correlation coefficient named r. When the points are found with positive slope, r will have positive value. When the points are negatively sloped, r will have negative value. The coefficient value determines how correct the setting is; this value ranges from -1 to 1. When the value of r is 1 or -1, it tells us that the points are adjusted exactly to the line. The closer r is to -1 or 1 indicates that there appears to be a linear relationship between the study variables. Otherwise when r is equal to 0, it tells us that the setting is not correct, and therefore it can be concluded that there is no apparent linear relationship.

The equation for determining the coefficient is as follows:

$$r = \frac{\text{cov}(x,y)}{\sigma_x \sigma_y},$$

where Cov represents the covariance of x, y. The symbols σ_x, σ_y represent the standard deviations of x and y.

Now we proceed to calculate the coefficient r for the created adjustment. For this we must introduce only the points of x and y, for the calculation of covariance and standard deviations.

```
In[58]:= r = N@ (Covariance@@{Data[[All,1]],Data[[All,2]]}) / (StandardDeviation@Data[[All,1]]*StandardDeviation@Data[[All,2]])
Out[58]= -0.987814
```

The result given to us is close to 1; therefore, we can say that the straight is adequately fair to the data. Although it is possible to calculate it through the equation, Mathematica has a function for this calculation. Correlation calculates the coefficient from two lists, so we need to enter only the x data in one list and the data from y in another list.

```
In[59]:= N@Correlation[Data[[All,1]],Data[[All,2]]]
Out[59]= -0.987814
```

And we get the same result as the previous one.

Linear Fit

Even using the aforementioned process, Mathematica has functions that specialize in finding the best linear model, with the use of LinearModelFit. Given the dataset, we write the LinearModelFit command with the data to work and the variable to write the equation. In addition, we can specify the level of precision for adjustment with WorkingPrecision.

```
In[60]:= Model=LinearModelFit[Data,x,x,WorkingPrecision→10]
Out[60]= FittedModel [8.473684211-1.466165414 x]
```

As we can see, the same equation returns to us but with better precision. Within the model we can access different properties related to the data, the model, and other adjustment parameters, and measures of the goodness of the fit, among others. To illustrate this, we see how to do it for the BestFit, BestFitParameters, and Function options, which are to return the equation of the best fit in the form of a list, the best parameters, and model construction for a pure function, respectively.

A very important aspect is that trying to make predictions about a future value using the fitted equation (8.47 - 1.47 x), with values of x outside the range could generate abnormal values, since we have not really established whether the relation of the equation outside the range of x is actually met. Figure 6-21 shows the fitted curve calculations.

```
In[61]:= {"\n"Framed["Best Fit Parameters b and m: "<>ToString[Model[
"BestFitParameters" ]],Background→LightYellow],"\n"Framed["Equation:
"<>ToString[Model["BestFit" ]],Background→LightYellow],
"\n"Framed["Pure Function:"<>ToString[SetPrecision[Model["Function"],3]],
Background→LightYellow],"\n"Framed["r coeficcient:"<>ToString[r],
Background→LightYellow]}
Out[61]=
```

```
{
  Best Fit Parameters b and m: {8.473684211, -1.466165414} ,
  Equation: 8.473684211 - 1.466165414 x ,
  Pure Function: 8.47 - 1.47 #1 & ,
  r coeficcient: -0.987814 }
```

Figure 6-21. *Fitted parameters, equation, and Pearson coefficient*

Since we have the line that best fits, we should consider whether there really is a relationship between x and y. How do you know if the adjustment made adequately describes the linear relationship between the variables x and y? To solve this problem, there is the concept of residual.

Model Properties

Residuals can be used as a measure to know how good the fit of the line is to the study points. Residuals are vertical deviations, either positive or negative. A residual point is the difference between the observed value of the dependent variable and the value that predicts the adjustment. To get the residual points we write the FitResiduals property within the model.

```
In[62]:= Model["FitResiduals"]
Out[62]= {0.06015038,0.52631579,-0.00751880,-0.54135338,-0.07518797,
0.39097744,-1.14285714,0.78947368}
```

With these points we can get the residual plot (Figure 6-22), which is the variable x vs the residual points.

```
In[63]:= ListPlot[Model["FitResiduals"],PlotStyle→{Red,Thick},PlotLabel→
"Residual Plot",AxesLabel→{Style["X",Bold],Style["residual points",
Bold]},Filling→Axis]
Out[63]=
```

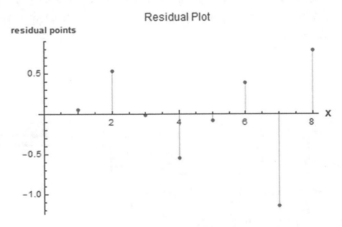

Figure 6-22. *Residual plot of the fitted data*

To show only the observed and predated values for the single prediction, use the SinglePredictionConfidenceIntervalTable option.

```
In[64]:= Model["SinglePredictionConfidenceIntervalTable"]
Out[64]=
```

Observed	Predicted	Standard Error	Confidence Interval
10	9.93984962	0.78481739	{8.0194706,11.8602286}
9	8.47368421	0.74856412	{6.6420138,10.3053546}
7	7.00751880	0.72287410	{5.2387096,8.7763280}
5	5.54135338	0.70889670	{3.8067456,7.2759611}
4	4.07518797	0.70732661	{2.3444221,5.8059538}
3	2.60902256	0.71824519	{0.8515399,4.3665052}
0	1.14285714	0.74110068	{-0.6705509,2.9562652}
-1	-1.78947368	0.81811053	{-3.7913180,0.2123707}

In addition to the residual points, we can extract the table from the parameters of the model adjusted with the ParameterTable property.

```
In[65]:= Model["ParameterTable"]
Out[65]=
```

	Estimate	Standard Error	t-Statistic	P-Value
1	8.473684211	0.34167121	24.800697	$2.8278226*10^{-7}$
-x	-1.466165414	0.094310214	-15.5461996	$1.4832546*10^{-6}$

The coefficients are shown in the table. The first coefficient is the ordinate to the origin, and the coefficient associated with the variable e is the slope. The two coefficients have their respective standard errors. To know the confidence interval for the parameters, we write the property ParameterConfidenceIntervalTable.

```
In[66]:= Model["ParameterConfidenceIntervalTable"]
Out[66]=
```

	Estimate	Standard Error	Confidence Interval
1	8.473684211	0.34167121	{7.63764488,9.30972355}
x	-1.466165414	0.094310214	{-1.69693419,-.23539663}

The default confidence interval is 95%. With these confidence values, we can plot the points that are inside or outside this range (Figure 6-23), extracting the values from the predictions and setting the option for the confidence interval to 0.95.

```
In[67]:= Model[x];
Model["SinglePredictionBands",ConfidenceLevel→0.95];
Show[ListPlot[Data,PlotStyle→Red],
Plot[{Model[x],Model["SinglePredictionBands",ConfidenceLevel→0.95]},
{x,-1,10},Filling→{2→{1}}],PlotRange→{Automatic,{-1,10}},Frame→True,
ImageSize→400]
Out[67]=
```

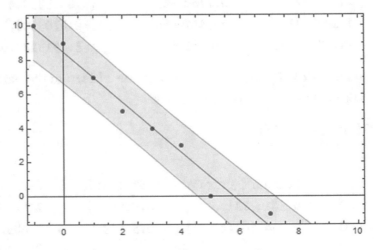

Figure 6-23. *The filled region denotes the 95% confidence interval.*

Finally, to obtain the properties related to the sum of the squared errors, we use the property of ANOVATable.

```
In[68]:= Model["ANOVATable"]
Out[68]=
```

	DF	SS	MS	F-Statistic	P-Value
x	1	107.213346	107.213346	241.68432	4.48325*10^-6
Error	6	2.6616541	0.44360902		
Total	7	109.8750000			

CHAPTER 7

Data Exploration

In this chapter, we will look at the basics of data management through the Wolfram Data Repository online platform. We will review how this website is built in order to have a better understanding of its use in Mathematica through the Wolfram Language. Examples will be carried out on how to download the data from this platform through the use of the Wolfram Language as well as its representation of data in the form dataset as well as using the Query command. We will also look at how data can be viewed inside datasets, how to apply user functions, and commands inside the format dataset.

Wolfram Data Repository

The Wolfram data repository is a website, which in turn is a repository of data, which is in the Cloud. This data repository contains information from different categories, such as computer science, meteorology, agriculture, sports, text and literature education, and many more. Although this repository belongs to Wolfram Research, it is characterized by being of public domain. In the Wolfram Data Repository, the information contained is computable data that has been selected, structured, and cured to be for direct use, to perform numerical calculations, estimates, analysis, statistics, or demonstrations, among others. The content hosted in this repository is data from many sources, globally known datasets, and publication data. All this information is designed so that any individual can access it globally. The Wolfram Data Repository system provides a data source that, in turn, also enables the storage of new information. The information that is stored in the repository is designed for direct implementation to the Wolfram Language.

As we saw in the data import section, we know whether the website is active by receiving an HTTP type response, as shown in Figure 7-1.

```
In[1]:= URLRead["https://datarepository.wolframcloud.com/"]
Out[1]=
```

J. Villalobos Alva, *Beginning Mathematica and Wolfram for Data Science*,
https://doi.org/10.1007/978-1-4842-6594-9_7

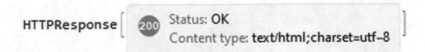

Figure 7-1. *Http response object of the Wolfram Data Repository. As we can see, we have received a successful response*

Wolfram Data Repository Website

To access this website, enter the following URL address in your favorite browser: `https://datarepository.wolframcloud.com`. Figure 7-2 shows the welcome page of the Wolfram Data Repository.

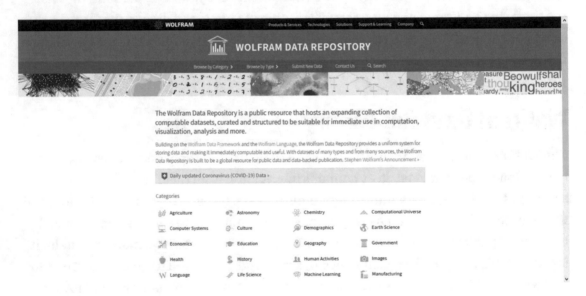

Figure 7-2. *Wolfram Data Repository website*

Note   The images that appear are links that redirect to the dataset associated with that image.

Once the site is loaded, we will see the repository title; below this, there is a menu of options to navigate the site, either by categories or by data type. Within that menu you will find the different categories that exist and the different types of data, be it text, numerical data, images, etc. Among the menu options, you will also find the contact option, custom searches, and Submit New Data. The latter is the option that redirects to another page that displays the instructions for publishing and uploading new data to this repository. If we scroll down, we will also see the categories that exist and the data types. If so, there is the possibility to browse all resources by clicking the Browse all resources link. To browse categories, we can choose the category from the menu or by clicking the name of the category at the initial site. Figure 7-3 shows what the site looks like once we have selected a category—in this case, Life Science.

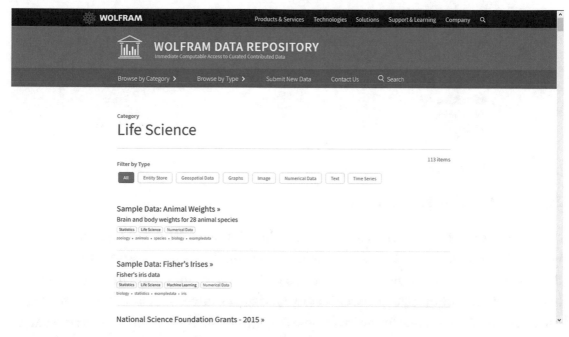

Figure 7-3. *Life Science category of the Wolfram Data Repository*

Note   The same process is for when we navigate by data type.

Selecting a Category

Each category shows the title, the number of elements contained in that category, and the option to filter the contents of that category by the type of data. Regarding the content, each sample data type is displayed with its title, a small description of the data it contains, and the different tags associated with that sample data. For example, the image shows Fisher's Irises known dataset. Once we select a sample dataset, it will take us to the site where the relevant information about that dataset is contained, as shown in Figure 7-4, where the Fisher's Irises dataset is selected.

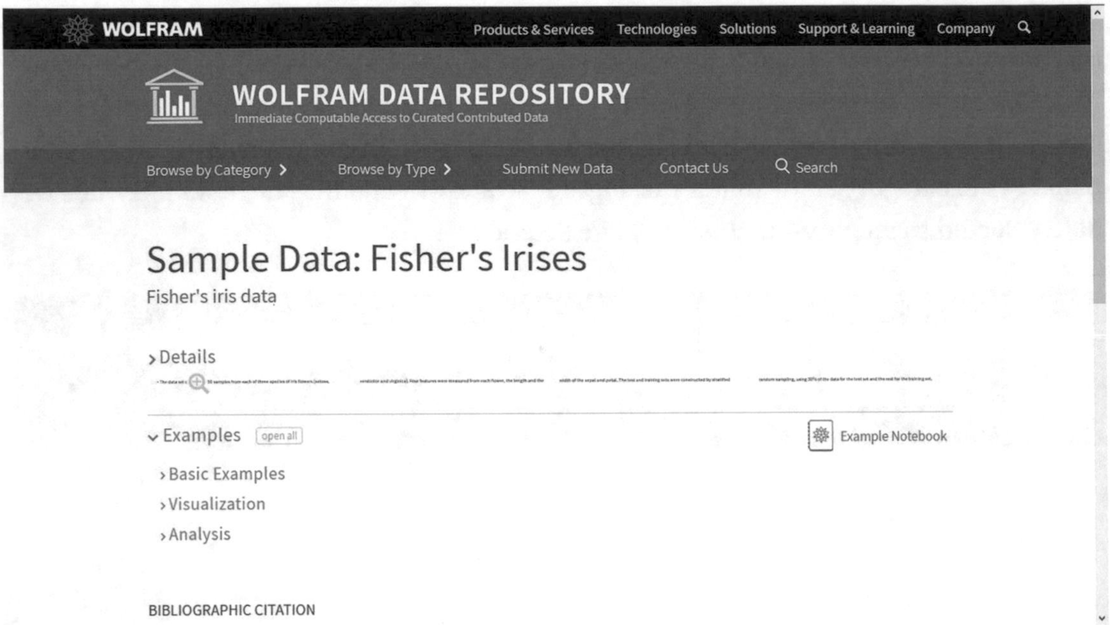

Figure 7-4. *Fisher's Irises dataset*

When a sample dataset is selected, a brief description of the dataset is shown as well as the different calculations that can be made and different formats to download the data or the notebook. Besides this, it also includes relevant information such as the bibliographic citation, data resource history, and data source. In certain cases, the data can either be downloaded for different types of formats such as comma-separated value (CSV), tab-separated value (TSV), JavaScript object notation (JSON), and others. Before starting to download data from the Wolfram Data Repository, it is necessary to have a Wolfram ID. This ID is an account that gives us access to the content of the Wolfram Data Repository in addition to other benefits such as the Wolfram One and Wolfram Alpha. To log in from Mathematica, head to the menu in Help ➤ Sign in, and a window will appear like the one in Figure 7-5.

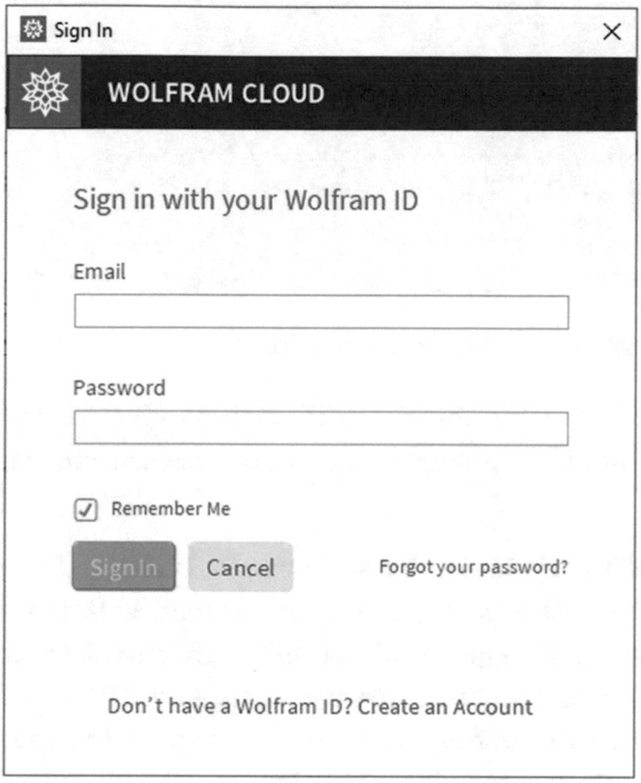

Figure 7-5. *Wolfram Cloud sign-in prompt*

In the new window, you will enter your email and password to be able to access from Mathematica the contents of the Wolfram Data Repository.

Extracting Data from the Wolfram Data Repository

Let's start by looking at the information and properties of the Fisher's dataset; for this we must retrieve the information through a ResourceObject. With ResourceObject (Figure 7-6) we can now view the different properties of the published data by clicking the plus icon. Detailed information about the data will display, such as sample name, type, version, size of the data, and many more.

```
In[2]:= ResourceObject["Sample Data: Fisher's Irises"]
Out[2]=
```

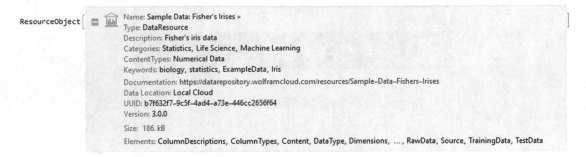

Figure 7-6. *ResourceObject of the Fisher´s Irises*

If we wanted to look to the properties of the resource object, enter the following code. This will give us a list of properties that can be accessed and that are related to the data sample.

```
In[3]:= ResourceObject["Sample Data: Fisher's Irises"]["Properties"]
Out[3]= {AllVersions,AutoUpdate,Categories,ContentElementLocations,ContentEle
ments,ContentSize,ContentTypes,ContributorInformation,DefaultContentElement,
Description,Details,Documentation,DocumentationLink,DOI,DownloadedVersion,
ExampleNotebook,ExampleNotebookObject,Format,InformationElements,Keywords,
LatestUpdate,Name,Originator,Properties,ReleaseDate,RepositoryLocation,
ResourceLocations,ResourceType,SeeAlso,ShortName,SourceMetadata,UUID,Version,
VersionInformation,VersionsAvailable,WolframLanguageVersionRequired}
```

Knowing already the list of properties related to information, we can now download from Mathematica the exercise notebook of the data sample.

```
In[4]:= ResourceObject["Sample Data: Fisher's Irises"]["ExampleNotebook"]
Out[4]= NotebookObject[Sample-Data-Fishers-Irises_examples.nb]
```

Once you finish evaluating the code, it will automatically open the new notebook. If we want to operate the notebook from the Cloud, we can type NotebookObject. This will give us back a Cloud, like object that is associated with a hyperlink.

```
In[5]:= ResourceObject["Sample Data: Fisher's Irises"]
["ExampleNotebookObject"]
Out[5]= CloudObject[https://www.wolframcloud.com/obj/5e59b79e-d95e-4f6f-
a7c8-f1276ba17be2]
```

If we press the link of the new notebook, it will open the internet browser and show us that it is in the Wolfram Cloud. Figure 7-7 shows this.

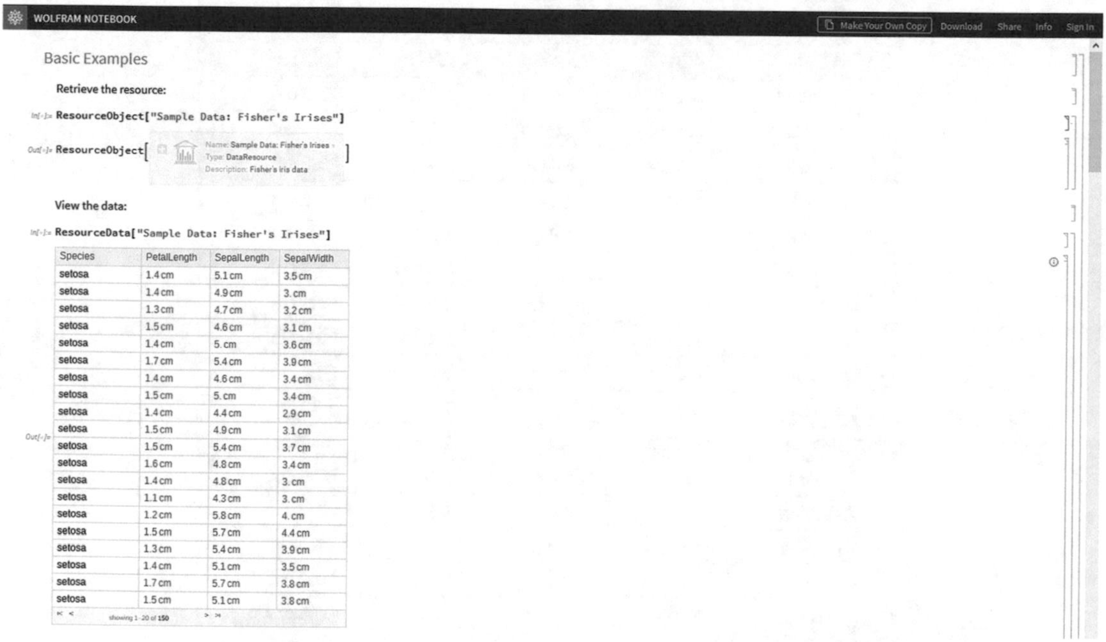

Figure 7-7. *Fisher´s Irises data sample, open from the Wolfram Cloud*

To access from Mathematica to the original sample data site, we enter Documentation, which will give us a URL object that you can enter to the site by clicking the double chevron icon.

```
In[6]:= ResourceObject["Sample Data: Fisher's Irises"]["Documentation"]
Out[6]= URL[https://datarepository.wolframcloud.com/resources/Sample-Data-
Fishers-Irises]
```

Accessing Data Inside Mathematica

The same initiative is applied to downloading the data using the ResourceData to the object resource. With ResourceData we access the contents of the specified resource; in this case, it is the Fisher's Irises data sample (Figure 7-8).

```
In[7]:= ResourceData[ResourceObject["Sample Data: Fisher's Irises"]]
Out[7]=
```

Species	SepalLength	SepalWidth	PetalLength	PetalWidth
setosa	5.1 cm	3.5 cm	1.4 cm	0.2 cm
setosa	4.9 cm	3. cm	1.4 cm	0.2 cm
setosa	4.7 cm	3.2 cm	1.3 cm	0.2 cm
setosa	4.6 cm	3.1 cm	1.5 cm	0.2 cm
setosa	5. cm	3.6 cm	1.4 cm	0.2 cm
setosa	5.4 cm	3.9 cm	1.7 cm	0.4 cm
setosa	4.6 cm	3.4 cm	1.4 cm	0.3 cm
setosa	5. cm	3.4 cm	1.5 cm	0.2 cm
setosa	4.4 cm	2.9 cm	1.4 cm	0.2 cm
setosa	4.9 cm	3.1 cm	1.5 cm	0.1 cm
setosa	5.4 cm	3.7 cm	1.5 cm	0.2 cm
setosa	4.8 cm	3.4 cm	1.6 cm	0.2 cm
setosa	4.8 cm	3. cm	1.4 cm	0.1 cm
setosa	4.3 cm	3. cm	1.1 cm	0.1 cm
setosa	5.8 cm	4. cm	1.2 cm	0.2 cm
setosa	5.7 cm	4.4 cm	1.5 cm	0.4 cm
setosa	5.4 cm	3.9 cm	1.3 cm	0.4 cm
setosa	5.1 cm	3.5 cm	1.4 cm	0.3 cm
setosa	5.7 cm	3.8 cm	1.7 cm	0.3 cm
setosa	5.1 cm	3.8 cm	1.5 cm	0.3 cm

rows 1–20 of 150

Figure 7-8. *Fisher's Irises dataset object*

As shown in Figure 7-8, the object that is returned is a ResourceData to use with a head of Dataset. Performing a visual inspection of the data sample, we observe that it is a dataset of 150 values, which contains five columns: Species, SepalLength, SepalWidth, PetalLength, and PetalWidth. If we pay attention, we can see how the values of the columns SepalLength, SepalWidth, PetalLength, and PetalWidth are quantities. Moving further down the entire dataset, we find that the species are divided into three categories: setosa, versicolor, and virginica. If we want to access the information related to the dataset, we must do it through the resource object and retrieve it through a ResourceData form, as shown.

```
In[8]:= ResourceObject["Sample Data: Fisher's Irises"]["ContentElements"]
Out[8]={ColumnDescriptions,ColumnTypes,Content,DataType,Dimensions,Observat
ionCount,RawData,Source,TrainingData,TestData}
```

With the ContentElements property, we are accessing the elements of the data sample, which are the ones that appear within the resource object. ContentElements shows us the information associated with the sample data, such as column information, data source, training data, test data, etc.—not to be confused with the properties of the resource object created, as it is not the same since you can construct a resource object for another associated name. To retrieve the information from the ContentElements, we must do it with ResourceData. This command will give us access to the contents of the data sample—in this case, the Fisher's Irises. Now let's get the data type of the columns.

```
In[9]:= ResourceData[ResourceObject["Sample Data: Fisher's
Irises"],"ColumnTypes"]
Out[9]= {Numeric,Numeric,Numeric,Numeric,Categorical}
```

The second argument of the ResourceData command is the element we are looking for. Running the aforementioned code shows us that there are four data types, three numeric and one categorical. Using a pure function, we can obtain information in a single expression. If we add the Column command, it is possible to have a better view of the information.

```
In[10]:= Column[ResourceData[ResourceObject["Sample Data: Fisher's
Irises"],#]&/@{"ColumnDescriptions","Dimensions","Source"}]
Out[10]= {Sepal length in cm.,Sepal width in cm.,Petal length in cm.,Petal
width in cm.,Species of iris}
{150,4}
Fisher,R.A. "The use of multiple measurements in taxonomic problems"
Annual Eugenics, 7, Part II, 179-188 (1936); also in "Contributions to
Mathematical Statistics" (John Wiley, NY, 1950).
```

This way we get to know the type of information in the columns ,such as dimensions, which are 150 rows per 4 columns and the data source.

Data Observation

In this part we will see how to observe data inside a dataset. We will use the Iris dataset, which has been extracted from the Wolfram Data Repository. Let's start by naming the data sample Fisher; this variable will contain the dataset with quantities included.

```
In[11]:= Fisher=ResourceData[ResourceObject["Sample Data: Fisher's Irises"]];
```

If we look at the dataset, we will notice that the numbers have their units and magnitude. Having the dataset, we can perform endless processes, such as grouping the content by the category variable that is the type of species. It is necessary to emphasize that I will access the dataset contained in the Fisher's variable. Let's look at the type of data that contains each column grouped by species (Figure 7-9).

```
In[12]:= Fisher[GroupBy["Species"]]
Out[12]=
```

	Species	SepalLength	SepalWidth	PetalLength	PetalWidth
setosa	setosa	5.1 cm	3.5 cm	1.4 cm	0.2 cm
	setosa	4.9 cm	3. cm	1.4 cm	0.2 cm
	setosa	4.7 cm	3.2 cm	1.3 cm	0.2 cm
	setosa	4.6 cm	3.1 cm	1.5 cm	0.2 cm
	setosa	5. cm	3.6 cm	1.4 cm	0.2 cm
	50 total ›				
versicolor	versicolor	7. cm	3.2 cm	4.7 cm	1.4 cm
	versicolor	6.4 cm	3.2 cm	4.5 cm	1.5 cm
	versicolor	6.9 cm	3.1 cm	4.9 cm	1.5 cm
	versicolor	5.5 cm	2.3 cm	4. cm	1.3 cm
	versicolor	6.5 cm	2.8 cm	4.6 cm	1.5 cm
	50 total ›				
virginica	virginica	6.3 cm	3.3 cm	6. cm	2.5 cm
	virginica	5.8 cm	2.7 cm	5.1 cm	1.9 cm
	virginica	7.1 cm	3. cm	5.9 cm	2.1 cm
	virginica	6.3 cm	2.9 cm	5.6 cm	1.8 cm
	virginica	6.5 cm	3. cm	5.8 cm	2.2 cm
	50 total ›				

Figure 7-9. *Iris data grouped by species*

Let us notice how the data is divided into three categories: setosa, versicolor, and virginica. If we pay attention to detail, we will notice that each of these categories contains a number 50 at the end of the Species column of each category. This means that there are 50 more rows in addition to those shown, making a total of 50 for each category that is 150 rows in total, which matches the number of 150 we review the dimensions of the sample data. In the meantime, if we click one of the categories, it will show us the columns for that category alone, as shown in the Figure 7-10. The same happens if we select the specific column within a category—it will show only that column for that category; try it to see what happens. There is also the possibility to click any column, and this will show us only the chosen column but for the three categories. By this I mean that if, for example, we choose SepalLength, we will see the contents of that column for the three species, as shown in Figure 7-10

⊞ › All › All › SepalLength					
setosa	5.1 cm	4.9 cm	4.7 cm	4.6 cm	5. cm
	5.4 cm	4.6 cm	5. cm	4.4 cm	4.9 cm
	5.4 cm	4.8 cm	4.8 cm	4.3 cm	5.8 cm
	5.7 cm	5.4 cm	5.1 cm	5.7 cm	5.1 cm
	5.4 cm	5.1 cm	4.6 cm	5.1 cm	4.8 cm
	50 total ›				
versicolor	7. cm	6.4 cm	6.9 cm	5.5 cm	6.5 cm
	5.7 cm	6.3 cm	4.9 cm	6.6 cm	5.2 cm
	5. cm	5.9 cm	6. cm	6.1 cm	5.6 cm
	6.7 cm	5.6 cm	5.8 cm	6.2 cm	5.6 cm
	5.9 cm	6.1 cm	6.3 cm	6.1 cm	6.4 cm
	50 total ›				
virginica	6.3 cm	5.8 cm	7.1 cm	6.3 cm	6.5 cm
	7.6 cm	4.9 cm	7.3 cm	6.7 cm	7.2 cm
	6.5 cm	6.4 cm	6.8 cm	5.7 cm	5.8 cm
	6.4 cm	6.5 cm	7.7 cm	7.7 cm	6. cm
	6.9 cm	5.6 cm	7.7 cm	6.3 cm	6.7 cm
	50 total ›				

Figure 7-10. *SepalLength column selected*

It is possible to group by species and choose only the columns that contain numeric values. This helps if, for example, we wanted to make a visual inspection of the dataset (Figure 7-11).

```
In[13]:= Query[GroupBy[Key["Species"]→KeyTake[{"SepalLength","SepalWidth",
"PetalLength","PetalWidth"}]]][Fisher]
Out[13]=
```

	SepalLength	SepalWidth	PetalLength	PetalWidth
setosa	5.1 cm	3.5 cm	1.4 cm	0.2 cm
	4.9 cm	3. cm	1.4 cm	0.2 cm
	4.7 cm	3.2 cm	1.3 cm	0.2 cm
	4.6 cm	3.1 cm	1.5 cm	0.2 cm
	5. cm	3.6 cm	1.4 cm	0.2 cm
	50 total ›			
versicolor	7. cm	3.2 cm	4.7 cm	1.4 cm
	6.4 cm	3.2 cm	4.5 cm	1.5 cm
	6.9 cm	3.1 cm	4.9 cm	1.5 cm
	5.5 cm	2.3 cm	4. cm	1.3 cm
	6.5 cm	2.8 cm	4.6 cm	1.5 cm
	50 total ›			
virginica	6.3 cm	3.3 cm	6. cm	2.5 cm
	5.8 cm	2.7 cm	5.1 cm	1.9 cm
	7.1 cm	3. cm	5.9 cm	2.1 cm
	6.3 cm	2.9 cm	5.6 cm	1.8 cm
	6.5 cm	3. cm	5.8 cm	2.2 cm
	50 total ›			

Figure 7-11. *Dataset with the species column suppressed*

What happens in the latter code is that we use the Key command to access the keys of the species column. Once these keys are accessed, we write a transformation rule so that each extracted key is assigned the associations extracted (KeyTake) from columns (SepalLength, SepalWidth, PetalLength, PetalWidth), then grouped and applied to Fisher's dataset.

If we wanted to count the data elements in the Fisher's dataset, we can add an ID column as a label (Figure 7-12) to list the data it contains. To achieve this, first we create an association with keys and values that go from 1 to the length of the dataset. Then this instruction is applied to the dataset object Fisher's, which adds the ID's as labels for the rows.

```
In[14]:= Query[AssociationThread[Range[Length@#]→Range[Length@#]]]
[Fisher]&[Fisher]
Out[14]=
```

	Species	SepalLength	SepalWidth	PetalLength	PetalWidth
1	setosa	5.1 cm	3.5 cm	1.4 cm	0.2 cm
2	setosa	4.9 cm	3. cm	1.4 cm	0.2 cm
3	setosa	4.7 cm	3.2 cm	1.3 cm	0.2 cm
4	setosa	4.6 cm	3.1 cm	1.5 cm	0.2 cm
5	setosa	5. cm	3.6 cm	1.4 cm	0.2 cm
6	setosa	5.4 cm	3.9 cm	1.7 cm	0.4 cm
7	setosa	4.6 cm	3.4 cm	1.4 cm	0.3 cm
8	setosa	5. cm	3.4 cm	1.5 cm	0.2 cm
9	setosa	4.4 cm	2.9 cm	1.4 cm	0.2 cm
10	setosa	4.9 cm	3.1 cm	1.5 cm	0.1 cm
11	setosa	5.4 cm	3.7 cm	1.5 cm	0.2 cm
12	setosa	4.8 cm	3.4 cm	1.6 cm	0.2 cm
13	setosa	4.8 cm	3. cm	1.4 cm	0.1 cm
14	setosa	4.3 cm	3. cm	1.1 cm	0.1 cm
15	setosa	5.8 cm	4. cm	1.2 cm	0.2 cm
16	setosa	5.7 cm	4.4 cm	1.5 cm	0.4 cm
17	setosa	5.4 cm	3.9 cm	1.3 cm	0.4 cm
18	setosa	5.1 cm	3.5 cm	1.4 cm	0.3 cm
19	setosa	5.7 cm	3.8 cm	1.7 cm	0.3 cm
20	setosa	5.1 cm	3.8 cm	1.5 cm	0.3 cm

rows 1–20 of 150

Figure 7-12. *ID's added to the Fisher´s dataset*

If we drag down the bar, we see that the counter reaches 150 elements.

In case you don't want to add an enumerated column to count the elements, we can use the Counts command (Figure 7-13).

```
In[15]:= Fisher[Counts,"Species"]
Out[15]=
```

setosa	50
versicolor	50
virginica	50

Figure 7-13. *Counted elements on the dataset*

This results in 50 data belonging to setosa, versicolor, and virginica. If we add them up, we get 150. You can also use the Query command, Query[Counts, "Species"] [Fisher].

Now let's see how to get the average of the three categories for each column. It would be possible if we knew the average of SepalLength, SepalWidth, PetalLength, and PetalWidth for the species, setosa, versicolor, and virginica, as exhibited in Figure 7-15.

```
In[16]:= Query[GroupBy[Key["Species"]→KeyTake[{"SepalLength","SepalWidth",
"PetalLength","PetalWidth"}]],Mean][Fisher]
Out[16]=
```

	SepalLength	SepalWidth	PetalLength	PetalWidth
setosa	5.006 cm	3.428 cm	1.462 cm	0.246 cm
versicolor	5.936 cm	2.77 cm	4.26 cm	1.326 cm
virginica	6.588 cm	2.974 cm	5.552 cm	2.026 cm

Figure 7-14. *Mean for the four columns, divided by species*

But, if we want to get the average of the columns for all categories, one way to get it would be by applying Mean as a query to the number of columns in the entire dataset (Figure 7-15).

```
In[17]:= Query[Mean][Fisher[[All,2;;5]]]
Out[17]=
```

SepalLength	5.84333 cm
SepalWidth	3.05733 cm
PetalLength	3.758 cm
PetalWidth	1.19933 cm

Figure 7-15. *Average values for the four columns of all species*

Note   The Mean command works with the quantities and returns the average to use as a quantity.

Descriptive Statistics

In this part we will see how to perform descriptive statistic of the Irises data and computations inside the format of the dataset as well as how to create custom grid formats. Let's get some descriptive statistics about this dataset. Let's create a function that would be called Stats. Let us start by building the function that will calculate the maximum, minimum, mean, median, first, and third quartile.

```
In[18]:= Stats[data_]:=
{
{#[{"Max:",Max@data}]},
{#[{"Min:",Min@data}]},
{#[{"Mean:",Mean@data}]},
{#[{"Median:",Median@data}]},
{#[{"1st quartile:",Quantile[data,0.25]}]},
{#[{"3rd quartile:",Quantile[data,0.75]}]}
}&[Row]
```

Now apply the created function to each of the columns. This is to get overall statistics for SepalLength, SepalWidth, PetalLength, and PetalWidth (Figure 7-16).

```
In[19]:= {{#1,#2,#3,#4},{Fisher[Stats,#1],Fisher[Stats,#2],Fisher[Stats,#
3],Fisher[Stats,#4]}}&["SepalLength","SepalWidth","PetalLength","PetalWid
th"]//Grid
Out[19]=
```

SepalLength	SepalWidth	PetalLength	PetalWidth
Max: 7.9 cm	Max: 4.4 cm	Max: 6.9 cm	Max: 2.5 cm
Min: 4.3 cm	Min: 2. cm	Min: 1. cm	Min: 0.1 cm
Mean: 5.84333 cm	Mean: 3.05733 cm	Mean: 3.758 cm	Mean: 1.19933 cm
Median: 5.8 cm	Median: 3. cm	Median: 4.35 cm	Median: 1.3 cm
1st quartile: 5.1 cm	1st quartile: 2.8 cm	1st quartile: 1.6 cm	1st quartile: 0.3 cm
3rd quartile: 6.4 cm	3rd quartile: 3.3 cm	3rd quartile: 5.1 cm	3rd quartile: 1.8 cm

Figure 7-16. *Function Stats applied to each column*

This also can be displayed in a compact form in a tab format with TabView (Figure 7-17).

```
In[20]:= TabView[{#1→Fisher[Stats,#1],#2→Fisher[Stats,#2],#3→Fisher[Stat
s,#3],#4→Fisher[Stats,#4]},ControlPlacement→Left]&["SepalLength","SepalWi
dth","PetalLength","PetalWidth"]
Out[20]=
```

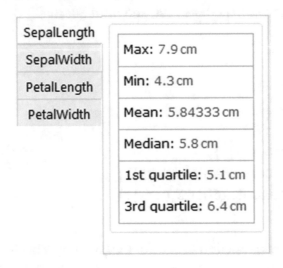

Figure 7-17. *Tabview format*

With TabView, we create three tabs with the names of each column, where it shows the values maximum, minimum, average, median, first, and third quartile; the columns are SepalLength, SepalWidth, PetalLength, and PetalWidth.

Table and Grid Formats

An alternative is to create a table for each species. In this way, we will create a better presentation of the data and thus be able to read it properly. We extract the data by applying the Nest command. With this command, we can specify the number of times a command or function will be applied; in this case, we will apply it twice.

```
In[21]:= Short[Values[Nest[Normal,Fisher,2]]]
{SLall,SWall,PLall,PWall}=%[[All,#]]&/@{2,3,4,5};
Out[21]//Short= {{setosa,5.1cm,3.5cm,1.4cm,0.2cm},{setosa,4.9cm,3.cm,1.4c
m,0.2cm},<<146>>,{virginica,6.2cm,3.4cm,5.4cm,2.3cm},{virginica,5.9cm,3.
cm,5.1cm,1.8cm}}
```

Having the values of all species separated by columns, we will proceed to create a list instead of a function, where the statistics will be displayed according to each column, adding calculations such as variance, standard deviation, skewness, and kurtosis. Then we will assign the calculations in the variable DescriptiveStats.

```
In[22]:= {Max[#],Min[#],Median[#],Mean[#],Variance[#],StandardDeviat
ion[#],Skewness[#],Kurtosis[#],Quantile[#,0.25],Quantile[#,.75]}&/@
{SLall,SWall,PLall,PWall};
DescriptiveStats=%;
```

A table (Figure 7-18) can be created with these calculations and adding the rows and column headings.

```
In[23]:=
TableHeads={
Style["Sepal Length",#1,ColorData["HTML"]["Maroon"],#2,#3],
Style["Sepal Width",#1,ColorData["HTML"]["YellowGreen"],#2,#3],
Style["Petal Length",#1,ColorData["HTML"]["SteelBlue"],#2,#3],
Style["Petal Width",#1,ColorData["HTML"]["Orange"],#2,#3]
}&["Title",Italic,20];
TableRows={
Style["Max",#1,#2],Style["Min",#1,#2],
Style["Median",#1,#2],Style["Mean",#1,#2],Style["Variance",#1,#2],
Style["Standard\n Deviation",#1,#2],
Style["Skewness",#1,#2],
Style["Kurtosis",#1,#2],
Style["1st quartile",#1,#2],
Style["3rd quartile",#1,#2]
}&["Text",Italic];
TableForm[DescriptiveStats,TableHeadings→{TableHeads,TableRows}]
Out[23]=
```

	Max	Min	Median	Mean	Variance	Standard Deviation	Skewness	Kurtosis	1st quartile	3rd quartile
Sepal Length	7.9 cm	4.3 cm	5.8 cm	5.84333 cm	0.685694 cm^2	0.828066 cm	0.311753	2.42643	5.1 cm	6.4 cm
Sepal Width	4.4 cm	2. cm	3. cm	3.05733 cm	0.189979 cm^2	0.435866 cm	0.315767	3.18098	2.8 cm	3.3 cm
Petal Length	6.9 cm	1. cm	4.35 cm	3.758 cm	3.11628 cm^2	1.7653 cm	−0.272128	1.60446	1.6 cm	5.1 cm
Petal Width	2.5 cm	0.1 cm	1.3 cm	1.19933 cm	0.581006 cm^2	0.762238 cm	−0.101934	1.66393	0.3 cm	1.8 cm

Figure 7-18. *Table showing descriptive statistics by the four features*

Note that the statistics are calculated with their units, with the exception of skewness and kurtosis, since by definition they are dimensionless. However, we can create a better structure from Grid because it is possible to add dividers like a spreadsheet format. To do this, we will add the TableRows to the data and then apply a transpose so that each calculated statistic is with its respective name. Subsequently we will add the column titles.

```
In[24]:= Transpose[Prepend[DescriptiveStats,TableRows]];
{" ",Style["Sepal Length",#1,ColorData["HTML"]
["Maroon"],#2,#3],Style["Sepal Width",#1,ColorData["HTML"]["YellowGreen"],#
2,#3],Style["Petal Length",#1,ColorData["HTML"]["SteelBlue"],#2,#3],
Style["Petal Width",#1,ColorData["HTML"]["Orange"],#2,#3]
}&["Title",Italic,20];
NewTable=Prepend[%%,%];
```

We proceed to create the table in the form of a spreadsheet (Figure 7-19).

```
In[25]:= Grid[NewTable,
ItemSize→{{None, Scaled[0.11], Scaled[0.11],Scaled[0.11]}},Background
→{{LightGray},None},Dividers→{{False},{1,2,3,4,5,6,7,8,9,10,11→True,-
2→Blue}},Alignment→Center]
Out[25]:=
```

	Sepal Length	*Sepal Width*	*Petal Length*	*Petal Width*
Max	7.9 cm	4.4 cm	6.9 cm	2.5 cm
Min	4.3 cm	2. cm	1. cm	0.1 cm
Median	5.8 cm	3. cm	4.35 cm	1.3 cm
Mean	5.84333 cm	3.05733 cm	3.758 cm	1.19933 cm
Variance	0.685694 cm^2	0.189979 cm^2	3.11628 cm^2	0.581006 cm^2
Standard Deviation	0.828066 cm	0.435866 cm	1.7653 cm	0.762238 cm
Skewness	0.311753	0.315767	-0.272128	-0.101934
Kurtosis	2.42643	3.18098	1.60446	1.66393
1st quartile	5.1 cm	2.8 cm	1.6 cm	0.3 cm
3rd quartile	6.4 cm	3.3 cm	5.1 cm	1.8 cm

Figure 7-19. *Grid view of the descriptive statistics*

260

If we want to build the table for each species, we will have to first separate the dataset by species with the Cases command. We use Cases since it gives us the freedom to work with patterns. First write the code to extract the raw data, and instead of using Short, we will use Shallow to suppress the 150 values.

```
In[26]:= Shallow[Values[Nest[Normal,Fisher,2]],1]
Out[26]//Shallow= {<<150>>}
```

We will create the table for the versicolor species and proceed to extract the values for versicolor and storethe values of the columns in the variables SLVersi, SWVersi, PLVersi, and PWVersi.

```
In[27]:= Shallow[Cases[%,{"versicolor",__}],1]
{SLVersi,SWVersi,PLVersi,PWVersi}=%[[All,#]]&/@{2,3,4,5};
Out[27]//Shallow= {<<50>>}
```

We do the same construction as before for the calculation of statistics. But instead of the white space, we add the name versicolor to distinguish that the table belongs to the versicolor specie.

```
In[28]:= TableRows;
{Max[#],Min[#],Median[#],Mean[#],Variance[#],StandardDeviation[#],Skewness[
#],Kurtosis[#],Quantile[#,0.25],Quantile[#,.75]}&/@{SLVersi,SWVersi,PLVersi
,PWVersi};
DescriptiveStats2=Prepend[%,%%];
Transpose[DescriptiveStats2];
{
Style["Versicolor","Text",Red,Italic,20],Style["Sepal
Length",#1,ColorData["HTML"]["Maroon"],#2,#3],
Style["Sepal Width",#1,ColorData["HTML"]["YellowGreen"],#2,#3],
Style["Petal Length",#1,ColorData["HTML"]["SteelBlue"],#2,#3],
Style["Petal Width",#1,ColorData["HTML"]["Orange"],#2,#3]
}&["Title",Italic,20];
NewTable2=Prepend[%%,%];
```

Now we build the table (Figure 7-20) for the species versicolor.

```
In[29]:= Grid[NewTable2,ItemSize→{{None, Scaled[0.11], Scaled[0.11],Scaled
[0.11]}},Background→{{LightGray},None},Dividers→{{False},{1,2,3,4,5,6,7,8
,9,10,11→True,-2→Blue}},Alignment→Center]
Out[29]=
```

Versicolor	Sepal Length	Sepal Width	Petal Length	Petal Width
Max	7. cm	3.4 cm	5.1 cm	1.8 cm
Min	4.9 cm	2. cm	3. cm	1. cm
Median	5.9 cm	2.8 cm	4.35 cm	1.3 cm
Mean	5.936 cm	2.77 cm	4.26 cm	1.326 cm
Variance	0.266433 cm^2	0.0984694 cm^2	0.220816 cm^2	0.0391061 cm^2
Standard Deviation	0.516171 cm	0.313798 cm	0.469911 cm	0.197753 cm
Skewness	0.10219	-0.351867	-0.588159	-0.0302363
Kurtosis	2.40117	2.55173	2.9256	2.51217
1st quartile	5.6 cm	2.5 cm	4. cm	1.2 cm
3rd quartile	6.3 cm	3. cm	4.6 cm	1.5 cm

Figure 7-20. *Descriptive stats for the versicolor specie*

We have only done this for the species of versicolor; if required the same process will be performed for each species. For example, if choose Cases with the other species, we would change the text to the corresponding specie.

Dataset Visualization

Having viewed the capabilities of the Wolfram Language to perform descriptive statistics within dataset, statistical charts can be implemented inside the dataset format, as we will see in this fragment.

We can have a better perspective from graphs, we will use the dataset format (Figure 7-21) to display the graphs by their species.

```
In[30]:= Fisher[GroupBy["Species"],DistributionChart[#,PlotTheme→"Classic",
PlotLabel→"PetalLength cm", GridLines→Automatic]&,"]&,"PetalLength"]
Out[30]=
```

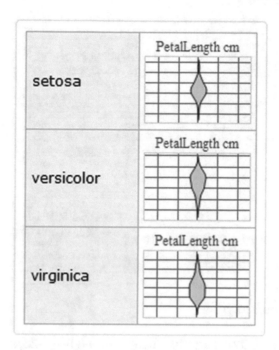

Figure 7-21. *Distribution chart plot*

We can perform the same process but for the box whiskers plot (Figure 7-22), but choose another column.

```
In[31]:= Fisher[GroupBy["Species"],BoxWhiskerChart[#,"Outliers",PlotTheme→
"Detailed",ChartLabels→Placed[{"SepalLength cm"},Above],BarOrigin→Right,C
hartStyle→Blue]&,"SepalLength"]
Out[31]=
```

Figure 7-22. _Box whiskers plot_

If the specie is clicked, it will amplify the graph (Figure 7-23).

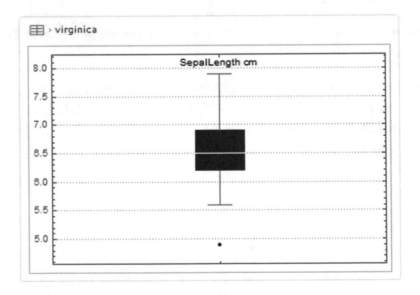

Figure 7-23. _Box whiskers plot for virginica specie_

The same applies for histograms. When the graph is very large, it appears suppressed within the dataset, but we can still select it, as shown in Figure 7-24.

```
In[32]:=Fisher[GroupBy["Species"],
 Labeled[Histogram[#,
    ColorFunction → (Hue[3/5, 2/3, #] &)], {Rotate["Frequency",
    90 Degree], "SepalWidth cm"}, {Left, Bottom}] &, "SepalWidth"]
Out[32]=
```

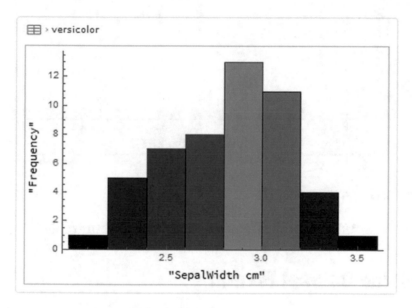

Figure 7-24. *Histogram plot for versicolor*

Here we show the 3D scatter plots for each species (Figure 7-25) for sepal length (x) vs sepal width (y).

```
In[33]:=Fisher[GroupBy["Species"],
 Labeled[ListPlot[{#, #}], {Rotate["Sepal width cm", 90 Degree], "Sepal
length cm"}, {Left, Bottom}] &, {"SepalLength","SepalWidth"}]
Out[33]=
```

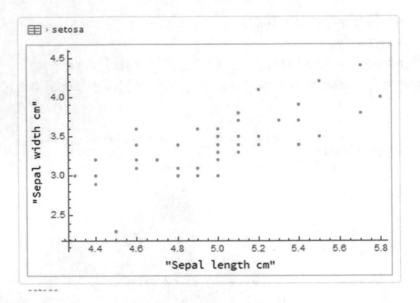

Figure 7-25. *2D scatter plot*

To return to the full dataset, click the dataset icon as with any other.

Data Outside Dataset Format

The truth is that there is also the possibility to extract the data crudely, as follows. We'll do this to have better data handling. We will use the Short command since the list is quite long.

```
In[34]:= Short[ResourceData[ResourceObject["Sample Data: Fisher's
Irises"],"RawData"]]
Out[34]//Short= {<<1>>}
```

With the data already extracted, we can get the values with the Values function and convert them to normal expressions.

```
In[35]:= Short[Normal[Values[%]]]
Out[35]//Short= {{setosa,5.1cm,3.5cm,1.4cm,0.2cm},{setosa,4.9cm,3.cm,1.4c
m,0.2cm},<<146>>,{virginica,6.2cm,3.4cm,5.4cm,2.3cm},{virginica,5.9cm,3.
cm,5.1cm,1.8cm}}
```

With the help of MapAt, we can extract the magnitudes of the quantities. The MapAt command gives us the freedom to choose where we want to apply the Quantity function. We choose to apply it to all rows with All, but only from column 2 to 4, which is where the quantities are located.

```
In[36]:= Short[Iris=MapAt[QuantityMagnitude,%,{All,2;;5}]]
Out[36]//Short= {{setosa,5.1,3.5,1.4,0.2},<<148>>,{virgini
ca,5.9,3.,5.1,1.8}}
```

It's worth asking a question here: Why do we remove the units if calculations can be made with them? We extract the magnitudes for all quantities because they have the same order of magnitude (cm), so each calculation will be in the same units, except if we made conversions or transformations to the data.

2D and 3D Plots

On the other hand, it is easier to manipulate lists with Wolfram Language. Having the data in the form of lists, we will now proceed to plot the three columns in a box plot and a distribution graph (Figure 7-26). We will only choose the three columns of the data.

```
In[37]:= Row[
{BoxWhiskerChart[{Iris[[All,#1]],Iris[[All,#2]],Iris[[All,#3]],Iris[[All,
#4]]},"Outliers",PlotRange→Automatic,FrameTicks→True,ChartStyle→"Sandy
Terrain",PlotLabel→"All Species",GridLines→Automatic,ChartLegends→Placed
[{"SepalLength","SepalWidth","PetalLength","PetalWidth"},Bottom],ImageSize
→Small],DistributionChart[{Iris[[All,#1]],Iris[[All,#2]],Iris[[All,#3]],
Iris[[All,#4]]},PlotRange→Automatic,FrameTicks→True,ChartStyle→"South
westColors",PlotLabel→"All Species",ChartLegends→Placed[{"SepalLength",
"SepalWidth","PetalLength","PetalWidth"},Bottom],PlotTheme→"Detailed",
GridLines→Automatic,ImageSize→Small]
}]&[2,3,4,5]
Out[37]=
```

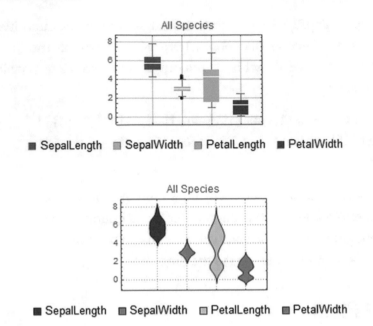

Figure 7-26. *Box whiskers plot and distribution chart for all species*

To improve this, let us graph for each species. We will use Cases to separate the list with their respective species (Figure 7-27).

```
In[38]:= Short[Setosa=Cases[Iris,{"setosa",__}]]
Short[Versi=Cases[Iris,{"versicolor",__}]]
Short[Virgin=Cases[Iris,{"virginica",__}]]
Out[38]//Short= {{setosa,5.1,3.5,1.4,0.2},<<48>>,{setosa,5.,3.3,1.4,0.2}}
Out[38]//Short= {{versicolor,7.,3.2,4.7,1.4},<<48>>,{versicol
or,5.7,2.8,4.1,1.3}}
Out[38]//Short= {{virginica,6.3,3.3,6.,2.5},<<48>>,{virgini
ca,5.9,3.,5.1,1.8}}
In[39]:= Column@{
BoxWhiskerChart[{Setosa[[All,#1]],Setosa[[All,#2]],Setosa[[All,#3]],Setosa
[[All,#4]]},"Outliers",PlotRange→Automatic,FrameTicks→True,ChartStyle→
"Rainbow",PlotLabel→"Setosa",ChartLegends→Placed[{"SepalLength","SepalWidth",
"PetalLength","PetalWidth"},Bottom],GridLines→Automatic],BoxWhiskerChart
[{Versi[[All,#1]],Versi[[All,#2]],Versi[[All,#3]],Versi[[All,#4]]},"Outliers",
PlotRange→Automatic,FrameTicks→True,ChartStyle→"Rainbow",PlotLabel→
"Versicolor",ChartLegends→Placed[{"SepalLength","SepalWidth","PetalLength",
"PetalWidth"},Bottom],GridLines→Automatic],BoxWhiskerChart[{Virgin[[All,#1]],
```

```
Virgin[[All,#2]],Virgin[[All,#3]],Virgin[[All,#4]]},"Outliers",PlotRange→
Automatic,FrameTicks→True,ChartStyle→"Rainbow",PlotLabel→"Virginica",
ChartLegends→Placed[{"SepalLength","SepalWidth","PetalLength","PetalWidth"},
Bottom],GridLines→Automatic]
}&[2,3,4,5]
Out[39]=
```

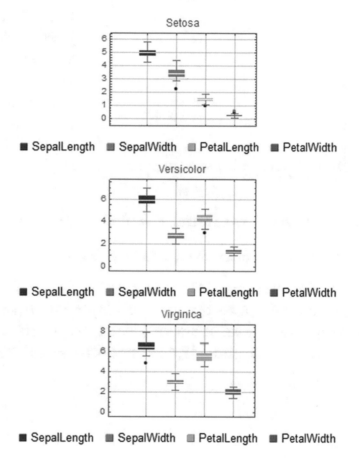

Figure 7-27. *Box whiskers plot for every specie with the four features*

In addition, we can join the scatter plots of sepal width vs sepal length for all species (Figure 7-28).

```
In[40]:= ListPlot[{Setosa[[All,{2,3}]],Versi[[All,{2,3}]],Virgin[[All,{2,3}
]]},FrameTicks→All,Frame→True,AspectRatio→1,PlotStyle→{Blue,Red,Green},
FrameLabel→{"Sepal length cm","Sepal width cm"},PlotLegends→{"Setosa",
"Versicolor","Virginica"}]
Out[40]=
```

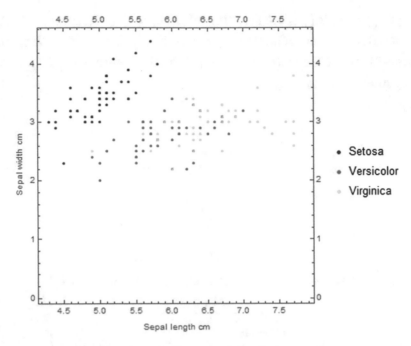

Figure 7-28. *2D scatter plot for all species of the first two features*

Or we can make a 3D scatter plot with three features (Figure 7-29).

```
In[41]:= ListPointPlot3D[{Setosa[[All,{2,3,4}]],Versi[[All,{2,3,4}]],Virgin
[[All,{2,3,4}]]},Ticks→All,AspectRatio→1,PlotStyle→{Blue,Red,Green},Axes
Label→{"Sepal length cm","Sepal width cm","Petal Length cm"},PlotLegends
→{"Setosa","Versicolor","Virginica"},PlotTheme→"Detailed",ViewPoint→{0,
-3, 3}]
Out[41]=
```

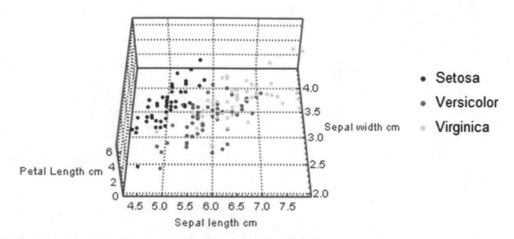

Figure 7-29. *3D scatter plot of three features for every species*

Now, when we have finished working with the resource object, we need to delete it so that the local cache of the resource is properly removed.

```
In[42]:=Clear[Fisher]
DeleteObject[ResourceObject["Sample Data: Fisher's Irises"]]
```

271

CHAPTER 8

Machine Learning with the Wolfram Language

The section will consist of the introduction of the gradient descent algorithm as an optimization method for linear regression; the corresponding computations will be shown as well as the concept of the learning curve of the model. Later, we will see how to use the specialized functions of the Wolfram Language for machine learning such as Predict, Classify and ClusterClassify, in the case of linear regression, for logistic regression and for cluster search. Adding to this, the different objects and results that these functions generate as well as the metrics to measure the model will be shown for these functions. In each case, we will explain which parts of the model are fundamental for the correct construction using the Wolfram Language. For this part of the book we will use examples of known datasets such as the Fisher's Irises dataset, Boston housing dataset, and the Titanic dataset.

Gradient Descent Algorithm

The gradient descent is an optimization algorithm that consists in finding the minimum of a function through an iterative process. To build the process, the squared error loss function is minimized with the linear model hypothesis of the shape $f(xl) = \theta 0 + \theta 1 * X_j$, around the point X_j. The loss function is given by the following expression.

$$J(\theta) = \frac{1}{2*N} * \sum_{j=1}^{N} \left(f\left(x_j\right) - y_j \right)^2$$

© Jalil Villalobos Alva 2021
J. Villalobos Alva, *Beginning Mathematica and Wolfram for Data Science*,
https://doi.org/10.1007/978-1-4842-6594-9_8

The iterative process of the algorithm consists of the calculation of the coefficients until convergence is obtained. The coefficients are given by the following expressions.

$$\theta_0^{i+1} = \theta_0^i - \frac{\alpha}{N} * \sum_{j=1}^{N} \left(\theta_0^i + \theta_1^i * x_j - y_j \right)$$

$$\theta_1^{i+1} = \theta_1^i - \frac{\alpha}{N} * \sum_{j=1}^{N} \left(\theta_0^i + \theta_1^i * x_j - y_j \right) * x_j$$

where the summations are obtained from partial derivatives with respect to $\theta 0$ and $\theta 1$. The term α corresponds to the learning rate, which is a parameter that minimizes error when the learning process is constructed. For more mathematical depth about the method and demonstrations, see the book *Artificial Intelligence: A Modern Approach* by Stuart Russell and Peter Norvig (2010, Upper Saddle River, NJ: Prentice Hall).

Getting the Data

To start we first define our data with the RandomReal function and establishing a seed. This is to maintain the reproducibility of the data, in case of practicing the same example.

```
In[1]:=
SeedRandom[888]
x=RandomReal[{0,1},50];
y=-1-x+0.6*RandomReal[{0,1},50];
```

Therefore, lets observe the data with a 2D scatter plot Figure 8-1.

```
In[2]:= ListPlot[Transpose[{x,y}],AxesLabel→{"X axis","Y axis"},
PlotStyle→Red]
Out[2]=
```

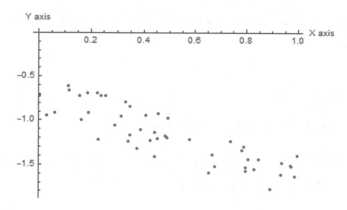

Figure 8-1. *2D scatter plot of the random generated data*

Algorithm Implementation

Let us now proceed to the implementation of the algorithm with the Wolfram Language. The algorithm will consist of defining the constants, the number of iterations, and the learning rate. Then we will create two lists containing initial values of zero, in which the values of the coefficients for each iteration will be stored. Later, we will perform the calculation of the coefficients through a loop with Table, which will not end until we reach the number of iterations. In our case, we will establish a number of iterations of 250 with a learning rate of 1.

```
In[3]:= itt=250;(*Number of iterations*)
α=1;(*Learning rate*)
θ0=Range@@{0,itt};(* Array for values of Theta_0*)
θ1=Range@@{0,itt};(* Array for values of Theta_1*)
Table[{
```
$$\theta0[[i+1]]=\theta0[[i]]- \frac{\alpha}{\text{Length}@x}x *\text{Sum}[(\theta0[[i]]+\theta1[[i]]* x[[j]]-y[[j]]),$$
```
{j,1,Length@x}];
```
$$\theta1[[i+1]]=\theta1[[i]]- \frac{\alpha}{\text{Length}@x}x *\text{Sum}[(\theta0[[i]]+\theta1[[i]]*x[[j]]-$$
```
y[[j]])*x[[j]],{j,1,Length@x}];},{i,1,itt}];
```

Since we have determined the calculation of the coefficients, we will build the linear adjustment equation by constructing a function and using the coefficient values of the last iteration, which are in the last position of the lists θ0 y θ1.

```
In[4]:= F[X_]:= θ0[[Length@ θ0]]+ θ1[[Length@ θ1]]*X
```

To know the shape of the best fit, we add the X variable as an argument. This will give us the form: $F(X) = \theta 0 + \theta 1{*}X$.

```
In[5]:= F[X]
Out[5]= -0.707789-0.923729 X
```

Let us look at how the line fits the data in Figure 8-2.

```
In[6]:= Show[{Plot[F[X],{X,0,1},PlotStyle→Blue,AxesLabel→{"X axis",
"Y axis"}],ListPlot[Transpose[{x,y}],PlotStyle→Red]}]
Out[6]=
```

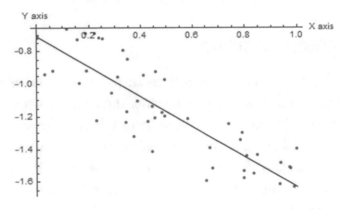

Figure 8-2.   *Adjusted line to the data*

Since we have built the linear model, we can make a graphical comparison of the variation of the learning rate with the number of iterations and the loss value given by the function J.

But first we must declare the loss function J. For the summation we can either use the special symbols of sigma or write $\sum_{i=1}^{imax} expr$ or Sum [expr, {i,i_{max}}].

```
In[7]:=J[Theta0_,Theta1_]:=1/(2*Length[x])*Sum[(Theta0 + (Theta1*x[[i]]) -
y[[i]])^2,{i,1,Length@x}]
```

Below is the graph of loss vs. each interaction for learning rate values of $\alpha 1=1$, $\alpha 2=0.1$, $\alpha 3=0.01$, $\alpha 4=0.001$, and $\alpha 5=0.001$, when repeating the process Figure 8-3.

Multiple Alphas

Having seen the previously constructed process, we can repeat the process for different alphas. Following is the graph of loss vs. each interaction for learning rate values of α1=1, α2=0.1, α3=0.01, α4=0.001, and α5=0.001, when repeating the process.

```
In[8]:=
α1=Transpose[{Range[0,itt], J[θ0,θ1]}];
α2=Transpose[{Range[0,itt], J[θ0,θ1]}];
α3=Transpose[{Range[0,itt], J[θ0,θ1]}];
α4=Transpose[{Range[0,itt], J[θ0,θ1]}];
α5=Transpose[{Range[0,itt], J[θ0,θ1]}];
```

Graph with ListLinePlot and visualize the learning curve for different alphas (Figure 8-3). When changing the value of alpha, try to check how also the adjusted line changes.

```
In[9]:= ListLinePlot[{α1,α2,α3,α4,α5},FrameLabel→{"Number of Iterations",
"Loss Function"},Frame→True,PlotLabel→"Learning Curve",PlotLegends→
SwatchLegend[{Style["α=1",#],Style["α=0.1",#],Style["α=0.01",#],Style[
"α=0.001",#],Style["α=0.0001",#]},LegendLabel→Style["Learning rate",White],
LegendFunction→(Framed[#,RoundingRadius→5,Background→Gray]&)]]&[White]
Out[9]=
```

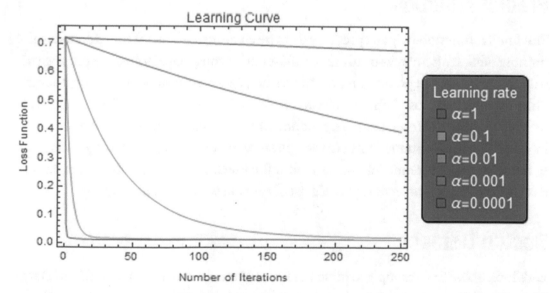

Figure 8-3. *Learning curve for the gradient descent algorithm*

In the previous graph (Figure 8-3) we can visualize the size of iterations with respect to cost and how it varies depending on the value of alpha. With a high learning rate, we can cover more ground at each step, but we risk exceeding the lowest point. To know whether the algorithm works, we must see that the loss function is decreasing in each new iteration. The opposite case would be an indicator that the algorithm is not working properly; this can be attributed to various factors such as a code error or an incorrect value of the learning rate. As we see in the graph, adequate values of alpha correspond to small values between a scale of 1 to 10^{-4}. It is not necessary to have to use these same values; you can use values that are within this range. Depending on the form of the data it is possible that the algorithm may or may not converge with different alpha values as the same for the iteration steps. If we choose very small alpha values, the algorithm can take a long time to converge, as we can see for alpha values 10^{-3} or 10^{-4}.

Linear Regression

Despite being able to build the algorithms to perform a linear regression, the Wolfram Language has a specialized function for machine learning. In the case of a linear regression problems, there is the Predict function. The Predict function can also work with different algorithms, not only regression task algorithms.

Predict Function

The Predict function helps us predict values by creating a predictor function using the training data. It also allows us to choose different learning algorithms, the purpose of which is to be able to predict a numerical, visual, categorical value or a combination. The methods to choose from are decision tree, gradient boosted tree, linear regression, neural network, nearest neighbors, random forest, and gaussian process. For each method, there are options within it; the options vary depending on the algorithm chosen to train the predictor function. Let us look at the linear regression method. The input data for Predict can be in the form of a list of rules, associations, or a dataset.

Boston Dataset

Let's look at the first example loading the Boston Homes data from the Wolfram Data Repository (Figure 8-4). The Boston dataset contains information about housing in the Boston Mass area. To look for more in-depth information, visit the article by David Harrison, and Daniel Rubinfeld, "Hedonic Housing Prices and the Demand for Clean Air" which appears in the. *Journal of Environmental Economics and Management*, (1978; 5[1], 81-102. https://doi.org/10.1016/0095-0696(78)90006-2) or the book *Regression Diagnostics: Identifying Influential Data and Sources of Collinearity: 546* by David Belsley, Edwin Kuh, and Roy Welsch, (2013; Wiley-Interscience).

```
In[1]:= Bstn=ResourceData[ResourceObject["Sample Data: Boston Homes"]]
Out[1]=
```

CRIM	ZN	INDUS	CHAS	NOX	RM	AGE	DIS	RAD	TAX
0.00632	18	2.31	tract does not bound Charles river	0.538 ppm	6.575	65.2	4.09	1	296
0.02731	0	7.07	tract does not bound Charles river	0.469 ppm	6.421	78.9	4.9671	2	242
0.02729	0	7.07	tract does not bound Charles river	0.469 ppm	7.185	61.1	4.9671	2	242
0.03237	0	2.18	tract does not bound Charles river	0.458 ppm	6.998	45.8	6.0622	3	222
0.06905	0	2.18	tract does not bound Charles river	0.458 ppm	7.147	54.2	6.0622	3	222
0.02985	0	2.18	tract does not bound Charles river	0.458 ppm	6.43	58.7	6.0622	3	222
0.08829	12.5	7.87	tract does not bound Charles river	0.524 ppm	6.012	66.6	5.5605	5	311
0.14455	12.5	7.87	tract does not bound Charles river	0.524 ppm	6.172	96.1	5.9505	5	311
0.21124	12.5	7.87	tract does not bound Charles river	0.524 ppm	5.631	100	6.0821	5	311
0.17004	12.5	7.87	tract does not bound Charles river	0.524 ppm	6.004	85.9	6.5921	5	311
0.22489	12.5	7.87	tract does not bound Charles river	0.524 ppm	6.377	94.3	6.3467	5	311
0.11747	12.5	7.87	tract does not bound Charles river	0.524 ppm	6.009	82.9	6.2267	5	311
0.09378	12.5	7.87	tract does not bound Charles river	0.524 ppm	5.889	39	5.4509	5	311
0.62976	0	8.14	tract does not bound Charles river	0.538 ppm	5.949	61.8	4.7075	4	307
0.63796	0	8.14	tract does not bound Charles river	0.538 ppm	6.096	84.5	4.4619	4	307
0.62739	0	8.14	tract does not bound Charles river	0.538 ppm	5.834	56.5	4.4986	4	307
1.05393	0	8.14	tract does not bound Charles river	0.538 ppm	5.935	29.3	4.4986	4	307
0.7842	0	8.14	tract does not bound Charles river	0.538 ppm	5.99	81.7	4.2579	4	307
0.80271	0	8.14	tract does not bound Charles river	0.538 ppm	5.456	36.6	3.7965	4	307
0.7258	0	8.14	tract does not bound Charles river	0.538 ppm	5.727	69.5	3.7965	4	307

rows 1-20 of 506   columns 1-10 of 14

Data not in notebook; Store now »

Figure 8-4. *Boston housing price dataset*

Try using the scroll bars to have a complete view of the dataset. Let's look at the descriptions of the columns and show them in TableForm.

```
In[2]:= ResourceData[ResourceObject["Sample Data: Boston Homes"],
"ColumnDescriptions"]//TableForm
Out[2]//TableForm=
Per capita crime rate by town
Proportion of residential land zoned for lots over 25000 square feet
```

Proportion of non-retail business acres per town
Charles River dummy variable (1 if tract bounds river, 0 otherwise)
Nitrogen oxide concentration (parts per 10 million)
Average number of rooms per dwelling
Proportion of owner-occupied units built prior to 1940
Weighted mean of distances to five Boston employment centers
Index of accessibility to radial highways
Full-value property-tax rater per $10000
Pupil-teacher ratio by town
1000(Bk-0.63)^2 where Bk is the proportion of Black or African-American
residents by town
Lower status of the population (percent)
Median value of owner-occupied homes in $1000s

Model Creation

We will try to create a model that is capable of predicting housing prices in the Boston
area through the number of rooms in the dwelling. To achieve this, the columns
of interest correspond to RM (average number of rooms per dwelling) and MEDV
(median value of owner-occupied homes), since we want to find out if there is a linear
relationship between the number of rooms and the price of the house. Applying a bit of
common sense, the houses with the largest number of rooms are larger and therefore
have the capacity to store more people, making the price go up.

Start by taking a look at the MEDV and RM scatter plot Figure 8-5.

```
In[3]:= MEDVvsRM=Transpose[{Normal[Bstn[All,"RM"]],Normal[Bstn[All,"MEDV"]]}];
ListPlot[MEDVvsRM,PlotMarkers→"OpenMarkers",Frame→True,FrameLabel→{Style[
"RM",Red],Style["MEDV",Red]},GridLines→All, PlotStyle→Black,ImageSize→500]
Out[3]=
```

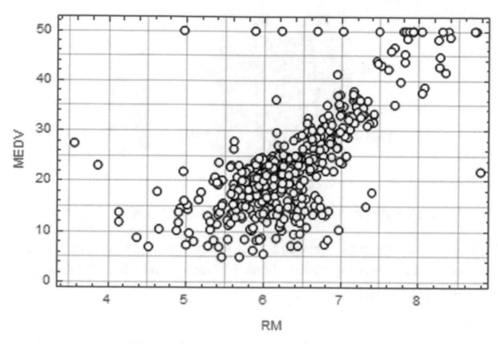

Figure 8-5. *2D scatter plot*

As seen in Figure 8-5, as the average number of rooms increases, the house price also increases. This suggests that there is possibly a directly proportional relationship between these two variables. Given what is seen in the graph, let us see the value of the correlation between these variables. We will show this through a correlation matrix, by first computing the correlation of the values, assigning the ticks names, and plotting it with MatrixPlot (Figure 8-6).

```
In[4]:= CorreLat=SetPrecision[Correlation[Transpose[{Normal[Bstn[All,"RM"]],
Normal[Bstn[All,"MEDV"]]}]],2];
XTicks={{1,"RM"},{2,"MEDV"},{1,"RM"},{2,"MEDV"}};
YTicks={{1,"RM"},{2,"MEDV"},{1,"RM"},{2,"MEDV"}};
PostionsValues={Text[#1,{0.5,1.5}],Text[#1,{1.5,0.5}],Text[#2,{1.5,1.5}],
Text[#2,{0.5,0.5}]}&[CorreLat[[1,1]],CorreLat[[1,2]]];
MatrixPlot[CorreLat,ColorFunction→"DarkRainbow",FrameTicks→{ XTicks,
YTicks,XTicks,YTicks},Epilog→{White,PostionsValues},PlotLegends→BarLegend
[{"DarkRainbow",{0,1}},4],ImageSize→180]
Out[4]=
```

By observing the matrix plot (Figure 8-6), it can be concluded that there is a good linear relationship between RM and MEDV.

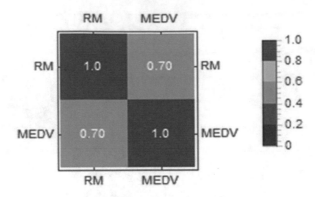

Figure 8-6. *Matrix plot combined with a correlation matrix*

Let's now shuffle the dataset randomly and establish a list of rules with Thread; this is because the data to be entered in the predictor function must be as follows: {x → y}—in other terms, input and target value.

```
In[5]:=
NewData=RandomSample[Thread[Normal[Bstn[All,"RM"]]→Normal[Bstn[All,
ssss"MEDV"]]]];
```

Once randomly sampled, we will select the first 354 elements (70%), this will be the training set and the rest 152 (30%) will be the test set.

```
In[6]:={training,test}={NewData[[;;354]],NewData[[355;;]]};
```

We proceed to train the model, a predictor for the average values of owner-occupied homes (MEDV) as a target. As a method we choose linear regression. When training a model, specification of the option of training report includes Panel (dynamical updating of the panel), Print (periodic information including time, training example, best method, current loss), ProgressIndicator (simple progress bar), SimplePanel (dynamic update panel with no plots), and None. Panel is the default option (Figure 8-7).

```
In[7]:=
PF=Predict[training,Method→"LinearRegression",TrainingProgressReporting→
"Panel"]
Out[7]=
```

Figure 8-7. *PredictorFunction object of the trained model*

When entering the code, depending on the option added to TrainingProgressReporting, a progress bar and panel report should appear (Figure 8-8). The time of the panel displayed depends on the training time of the model. To set a specific time for the training time, add as an option TimeGoal, which specifies how long the training should last for the model. Time values are seconds of CPU time—that is, the number with no units. With units of time (seconds, minutes, and hours), the use of Quantity command is needed, like TimeGoal → Quantity ["time magnitude", #] & / @ {"Second", "Minute", "Hour"}.

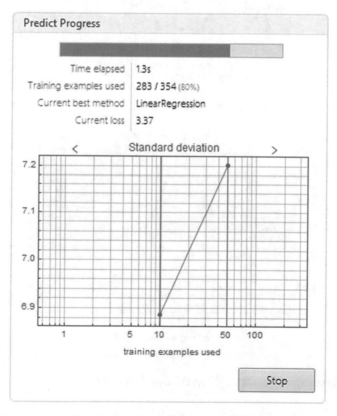

Figure 8-8. *Progress report of the PredictorFunction*

Back to the model: as seen in Figure 8-7, the return object is a predictor function (try using Head to verify it). When having assigned a name to the predictor function, additional information about the model can be obtained for this; the command Information is used (Figure 8-9). Information works for every other expression, not just for machine learning purposes.

```
In[8]:= Information[PF]
Out[8]=
```

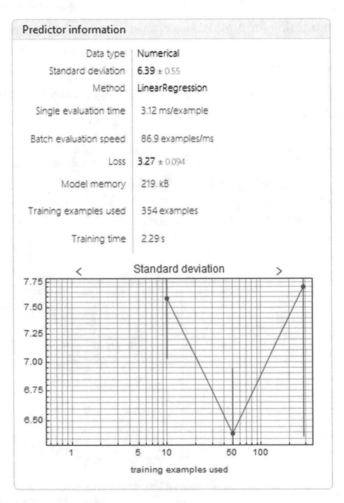

Figure 8-9. *Information report of the trained model*

The information panel (Figure 8-9) includes data type, root mean squared (StandardDeviation), method, batch evaluation speed, loss, model memory, number of examples for training, and training time. The graphics at the bottom of the panel are for standard deviation, model learning curve, and learning curve for the other algorithms. If you hover the cursor pointer over the numerical parameters, it will show the confidence intervals and units. If it's done by the name of the method, it will show the parameters of the linear regression method. Since we did not select a specific optimization algorithm within the LinearRegression method, Mathematica tries to search through the algorithms for the best one (this can be viewed in the learning curve for all algorithms). We will see how to access these options further down the line.

Note   Every method that can be used in the Predict function has options and suboptions; to see full customization use the Wolfram Language Documentation Center.

Table 8-1 shows the different common options that can be used for model training, as well as their definition and possible values for the training process of a PredictorFunction.

Table 8-1. *Most Common Options for Predict Function*

Option	Definition
Method	Algorithm Possible values: DecisionTree, GradientBoostedTrees, LinearRegression, NearestNeighbors. NeuralNetwork, RandomForest and GaussianProcess.
PerformanceGoal	Performance optimization Possible values: DirectTraining, Memory, Quality, Speed, TrainingSpeed, Automatic. Combination of values is supported (P PerformanceGoal→ {val1, val2}).
RandomSeeding	Seed for the pseudorandom number generator Possiblevalues: Automatic. "custom seed," Inherited (random seed used in previous computations).
TargetDevice	Specify a device to perform the training or test process Possible values: CPU or GPU. If a GPU is installed, the automatic target device will be the GPU:
TimeGoal	Time spent for the training process
TrainingProgressIndicator	Progress report Possible values: Panel, Print, ProgressIndicator, SimplePanel, None.

Model Measurements

Once the model is built, we must observe and analyze the performance of the predictor function in the test set. To carry out this, we must do it within the PredictorMeasurments command. The predictor function goes in the argument (Figure 8-10), followed by the test set, followed by the property or properties to add.

```
In[9]:= PRM=PredictorMeasurements[PF,test]
Out[9]=
```

Figure 8-10. *PredictorMeasurements object of the tested model*

The returned object is called PredictorMeasurementsObject (Figure 8-10). We can add the properties from the PredictorMeasurements command. We can assign a variable to the object to access it more simply. Let's look at the model report with the test set (Figure 8-11).

```
In[10]:= PRM["Report"]
Out[10]=
```

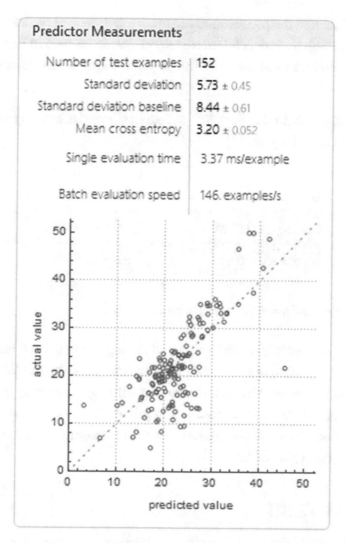

Figure 8-11. *Report of tested model*

This report (Figure 8-11) shows different parameters, such as the root mean square (standard deviation), mean cross entropy, among others. And it shows us a graph of the fit of the model along with the current values and predicted values. We see that the model is good for most cases, with the exception that there are still some outliers that affect performance.

To better understand the precision of the model, let's look at the root mean squared error (RMSE) and RSquared (coefficient of determination) shown in Figure 8-12. To display the associated uncertainties, use the option ComputeUncertainty with true value.

```
In[11]:= Dataset[AssociationMap[PRM[#,ComputeUncertainty→True]&,{
"StandardDeviation","RSquared"}]]
Out[11]=
```

StandardDeviation	5.7 ± 0.4
RSquared	0.54 ± 0.10

Figure 8-12. *Standard deviation and r-squared values*

This gives us a slightly high RMSE value and not a good r-squared value. Remembering that the value of r squared indicates how good the model is for making predictions. These two values would indicate that although there may be a linear relationship between the number of rooms and prices, this is not necessarily explained by a linear regression. These observations are also consistent, remembering that we obtained a correlation value of 0.7.

Model Assessment

The graphs made within the model are the model graph and the target variable (ComparisonPlot). To check the distribution of the variance, use the ResidualHistogram function, and to check the residual plot, use ResidualPlot. These are shown in Figure 8-13.

```
In[12]:= PRM[#]&/@{"ResidualHistogram","ResidualPlot","ComparisonPlot"}
Out[12]=
```

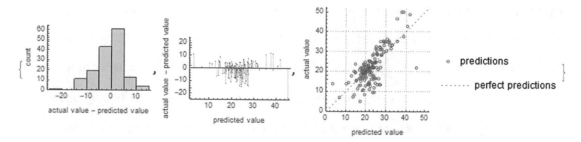

Figure 8-13. *ResidualHistogram, ResidualPlot, and ComparisonPlot*

To find out all the properties of the Predictor Measurements object, we write Properties as an argument. These properties can vary between methods.

```
In[13]:= PRM["Properties"]
Out[13]= {BatchEvaluationTime,BestPredictedExamples,ComparisonPlot,
EvaluationTime,Examples,FractionVarianceUnexplained,GeometricMeanProbability
Density,LeastCertainExamples,Likelihood,LogLikelihood,MeanCrossEntropy,
MeanDeviation,MeanSquare,MostCertainExamples,Perplexity,PredictorFunction,
ProbabilityDensities,ProbabilityDensityHistogram,Properties,RejectionRate,
Report,ResidualHistogram,ResidualPlot,Residuals,RSquared,StandardDeviation,
StandardDeviationBaseline,TotalSquare,WorstPredictedExamples}
```

In the event that we are not satisfied with the chosen methods or hyperparameters, retraining the model can be done by configuring the new values for the hyperparameters. We access the values of the current method with the help of the Information command and adding the properties of Method (shows us the method used to train the model), MethodDescription (description of the method used), and MethodOption (method options).

```
In[14]:= Information[PF,"MethodOption"]
Out[14]= Method→{LinearRegression,L1Regularization→0,L2Regularization→
0.00001,OptimizationMethod→NormalEquation}
```

As we can see, there are terms such as L1Regularization, L2Regularization, and OptimizationMethod. The first two terms are associated with regularization methods, and L1 refers to the Lasso regression name and L2 to the Ridge regression name. Regularization is used to minimize the complexity of the model, in addition to reducing the variation; it also improves the precision of the model, solving problems of overfitting.

This is accomplished by adding a penalty to the loss function; this penalty is added to the sum of the absolute value of the coefficient $\lambda_1 * \sum_{i=0}^{N} |\theta_i|$, whereas for L2, it is given by the expression $(\lambda_2 / 2) * \sum_{i=0}^{N} \theta_i^2$, where the function to minimize is the loss function $(1/2) * \sum_{i=0}^{N} (y_i - f(\theta, x_i))^2$. For more mathematical depth, visit *Artificial Intelligence: A Modern Approach.* by Stuart Russell and Peter Norvig (2010 Upper Saddle River, NJ: Prentice Hall) and *An Introduction to Statistical Learning: With Applications in R* by Gareth James, Trevor Hastie, Robert Tibshirani, and Daniela Witten (2017; 1st ed. 2013, Corr. 7th printing 2017 ed.: Springer). The third term is the option of which optimization method we want to choose; the existing methods are NormalEquation, StochasticGradientDescent, and OrthantWiseQuasiNewton. That said, it must be emphasized that when using the vector of coefficients with the L1 and L2 standards, this is known as an Elastic Net regression model. Elastic Net might be used in circumstantces when there is correlation in the parameters. For more theory, use the next reference, *The Elements of Statistical Learning: Data Mining, Inference, and Prediction, Second Edition* by Trevor Hastie, Robert Tibshirani, and Jerome Friedman(2nd 2009, Corr. 9th Printing 2017 ed.: Springer).

Retraining Model Hyperparameters

As discussed later, let's retrain the model but with the values of L1 → 12, L2 → 100 and the optimization algorithm OptimizationMethod → StochasticGradientDescent, TrainingProgressReporting → None, PerformanceGoal → "Quality", RandomSeeding → 10000, TargetDevice → "CPU".

```
In[15]:= PF2=Predict[training,Method→{"LinearRegression","L1Regularization"→
12,"L2Regularization"→100,"OptimizationMethod"→ Automatic},TrainingProgress
Reporting→None,PerformanceGoal→"Quality",RandomSeeding→10000,TargetDevice→
"CPU"];
```

To see the properties related to an example, type properties after the input data for the Predictorfunction—for example, PF2["example", "Properties"].

Now, let's compare the performance of the new model by showing the graphs and metrics like before (Figure 8-14 and Figure 8-15).

```
In[16]:= PRM2=PredictorMeasurements[PF2,test];
PRM[#]&/@{"ResidualHistogram","ResidualPlot","ComparisonPlot"}
Dataset[AssociationMap[PRM2[#,ComputeUncertainty→True]&,{
"StandardDeviation","RSquared"}]]
Out[16]=

Out[16]=
```

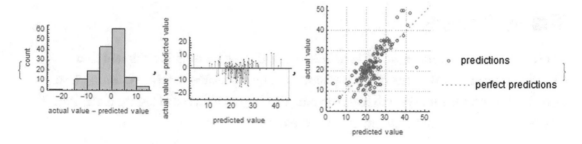

Figure 8-14. *Plots of the retrained model*

StandardDeviation	5.9 ±0.4
RSquared	0.51 ±0.10

Figure 8-15. *New values for RMSE and r squared*

Making observations in the graphs, we see the model merely decrease to a certain degree; this agrees with the new value of r squared, which decreases to 0.51. However, it is still a poor model when it comes to making future predictions. This can be attributed to the optimization choice, the L1 and L2 parameters choice.

Logistic Regression

Logistic regression is a technique commonly used in statistics, but it is also used within machine learning. The logistic regression works considering that the values of the response variable only take two values, 0 and 1; this can also be interpreted as a false or true condition. It is a binary classifier that uses a function to predict the probability of whether or not a condition is met, depending on how the model is constructed. Usually, this type of model is used for classification, since it has the ability to provide us with probabilities and classifications, since the values of the logistic regression oscillates between two values. In logistic regression, the target variable is a binary variable that contains encoded data. For further view visit *Introduction to Data Science: A Python Approach to Concepts, Techniques and Applications* by Laura Igual, Santi Seguí, Jordi Vitrià, Eloi Puertas, Petia Radeva, Oriol Pujol, Sergio Escalera, Francesc Dantí, and Lluis Garrido (2017 ed.: Springer).

Titanic Dataset

For the following example we will use the titanic dataset, which is a dataset that describes the survival status of the passengers. The variables used are class, age, sex, and survival condition. We will load the data directly as a dataset (Figure 8-16) from the ExampleData and enumerate the rows of the dataset.

Note   This section will be entirely constructed with the use of Query language so that the reader can understand how to use it more deeply inside datasets.

```
In[1]:= Titanic=Query[AssociationThread[Range[Length@#]→Range[Length@#]]]
[ExampleData[{"Dataset","Titanic"}]]&[ExampleData[{"Dataset","Titanic"}]]
Out[1]=
```

	class	age	sex	survived
1	1st	29	female	True
2	1st	1	male	True
3	1st	2	female	False
4	1st	30	male	False
5	1st	25	female	False
6	1st	48	male	True
7	1st	63	female	True
8	1st	39	male	False
9	1st	53	female	True
10	1st	71	male	False
11	1st	47	male	False
12	1st	18	female	True
13	1st	24	female	True
14	1st	26	female	True
15	1st	80	male	True
16	1st	—	male	False
17	1st	24	male	False
18	1st	50	female	True
19	1st	32	female	True
20	1st	36	male	False

rows 1–20 of 1309

Figure 8-16. *Titanic dataset*

Let's look at the dimensions of the data using the Dimensions command.

```
In[2]:= Dimensions@Titanic
Out[2]= {1309,4}
```

Interpreting the result, we see that the dataset comprises 1309 rows by 4 columns. Looking at the dataset, there are four columns classified by class, age, sex, and survived status. If we use the space bar we see that there are some elements that do not register data entry. To see which columns contain missing data, execute the following code by counting the number of elements that correspond to the pattern Missing in each of the columns.

```
In[3]:= Query[Count[_Missing],#]@Titanic&/@{"class","age","sex","survived"}
Out[3]= {0,263,0,0}
```

This gives us as a result that there are 263 missing values within the age column and zero for the others. Let's remove the rows that contain this missing data, but first we will extract the row numbers from the missing data by selecting the elements from the age column that are equal to Missing, then extracting the row IDs.

```
In[4]:= Query[Select[#age==Missing[]&]][Titanic];
Normal@Keys@%
Out[5]= {16,38,41,47,60,70,71,75,81,107,108,109,119,122,126,135,148,153,158
,167,177,180,185,197,205,220,224,236,238,242,255,257,270,278,284,294,298,31
9,321,36,4,383,385,411,470,474,478,484,492,496,525,529,532,582,596,598,673,
681,682,683,706,707,757,758,768,769,776,790,796,799,801,802,803,805,806,809
,813,814,816,817,820,836,843,844,853,855,857,859,866,872,873,875,877,880,88
3,887,888,901,902,903,904,919,921,922,923,924,927,928,929,930,931,932,941,9
43,945,946,947,949,955,956,957,958,959,962,963,972,974,977,983,984,985,988,
989,990,992,994,995,998,999,1000,1001,1002,1003,1004,1005,1006,1007,1010,10
13,1014,1015,1017,1019,1023,1024,1028,1029,1030,1031,1033,1034,1035,1036,10
37,1038,1039,1040,1042,1043,1044,1045,1053,1054,1055,1056,1070,1071,1072,10
73,1074,1075,1077,1078,1079,1081,1082,1086,1096,1110,1115,1116,1117,1122,11
23,1124,1125,1129,1133,1136,1137,1138,1139,1150,1151,1152,1155,1156,1160,11
63,1164,1165,1167,1168,1169,1171,1173,1174,1175,1176,1177,1178,1179,1180,11
81,1185,1186,1187,1194,1195,1196,1198,1199,1200,1201,1203,1213,1214,1215,12
16,1217,1220,1222,1242,1243,1244,1246,1247,1248,1250,1251,1254,1256,1263,12
69,1283,1284,1285,1292,1293,1294,1298,1303,1304,1306}
```

These numbers represent the rows that contain the missing data for the age column. To eliminate them we use the DeleteMissing command, considering that there is missing data at level 1. The final dataset is seen in (Figure 8-17)

```
In[5]:= Titanic=DeleteMissing[Titanic,1,1]
Out[5]=
```

	class	age	sex	survived
1	1st	29	female	True
2	1st	1	male	True
3	1st	2	female	False
4	1st	30	male	False
5	1st	25	female	False
6	1st	48	male	True
7	1st	63	female	True
8	1st	39	male	False
9	1st	53	female	True
10	1st	71	male	False
11	1st	47	male	False
12	1st	18	female	True
13	1st	24	female	True
14	1st	26	female	True
15	1st	80	male	True
17	1st	24	male	False
18	1st	50	female	True
19	1st	32	female	True
20	1st	36	male	False
21	1st	37	male	True

rows 1–20 of 1046

Figure 8-17. *Titanic dataset without missing values*

To corroborate that there is no longer any missing data, you could apply the same code with counts or by looking at the keys of the removed rows, for example.

```
In[6]:= Titanic[Key[16]]
Out[6]= Missing[KeyAbsent,16]
```

This means that there is no content associated with key 16. If you want to check all keys, use the row list of the missing data.

Data Exploration

Once we have removed the missing data, we can count the number of elements that consist of each class, sex, and survival status (Figure 8-18).

```
In[7]:= Dataset@
<|
"Class"→Query[Counts,"class"]@Titanic,"Sex"→ Query[Counts,"sex"]@Titanic,
"Survival status"→Query[Counts,"survived"]@Titanic
|>
Out[7]=
```

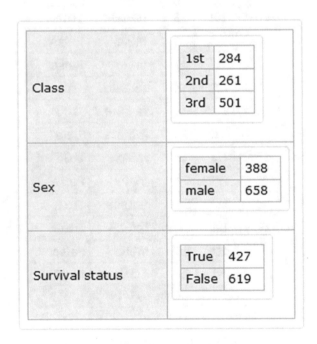

Figure 8-18. *Basic elements count for class, sex, and survival status*

After eliminating the rows with the missing elements, we see that the dataset consists of 284 elements in first class, 261 in second class, and 501 in third class (Figure 8-19). Also note that more than half of the registered passengers were male and that there were more deaths than survivors. It is possible to verify this graphically by showing the percentages (Figure 8-19). The same approach is applied to the columns class and sex.

```
In[8]:= Row[{PieChart[{N@(#[[1]]/Total@#),N@(#[[2]]/Total@#)}&[Counts
[Query[All,"survived"][Titanic]]], PlotLabel→Style["Percentage of
survival",#3,#4], ChartLegends→ {"Survived", "Died"}, ImageSize→#1,ChartS
tyle→#2,LabelingFunction→(Placed[Row[{SetPrecision[100#,3],"%"}],"RadialC
allout"]&)],
PieChart[{N@(#[[1]]/Total@#),N@(#[[2]]/Total@#)}&[Counts[Query[All,
"sex"][Titanic]]], PlotLabel→Style["Percentage by sex",#3,#4],
ChartLegends→{"Female", "Male"}, ImageSize→#1,ChartStyle→#2,LabelingFunc
tion→(Placed[Row[{SetPrecision[100#,3],"%"}],"RadialCallout"]&)],
PieChart[{N@(#[[1]]/Total@#),N@(#[[2]]/Total@#),N@(#[[3]]/Total@#)}&[Counts
[Query[All,"class"][Titanic]]], PlotLabel→Style["Percentage by class",#3,#4],
ChartLegends→{"1st", "2nd","3rd"}, ImageSize→#1,ChartStyle→#2,Labeling
Function→(Placed[Row[{SetPrecision[100#,3],"%"}],"RadialCallout"]&)]},"----
"]&[200,{ColorData[97,20],ColorData[97,13],ColorData[97,32]},Black,20]
Out[8]=
```

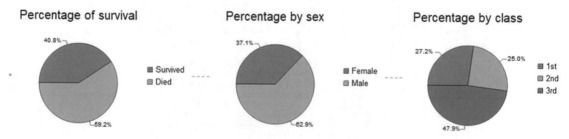

Figure 8-19. *Pie charts for class, sex, and survival status*

For this case, we are going to predict the survival of the Titanic passengers. We will build a model that will classify whether the given class, age, and sex will survive or not. The features will be class, age, and sex, and the target will be the survival status. We are going to use these variables as the features, which the model will then use to classify whether their class, age, and sex survive or not, which is our target variable. For this we divide the dataset into 80% training (837 elements) and 20% test (209 elements). To split the dataset, first we will do a random sampling; after we will extract the keys of the IDs and create the new dataset divided by train and test set (Figure 8-20).

```
In[9]:= BlockRandom[SeedRandom[8888];
RandomSample[Titanic]];
Keys@Normal@Query[All][%];
```

```
{train,test}={%[[1;;837]],%[[838;;1046]]};
dataset=Query[<|"Train"→{Map[Key,train]},"Test"→{Map[Key,test]} |> ]
[Titanic]
Out[9]=
```

		class	age	sex	survived
Train	410	2nd	36	male	False
	537	2nd	32	female	True
	874	3rd	42	male	False
	691	3rd	22	male	False
	1021	3rd	21	male	False
	852	3rd	45	female	True
	705	3rd	21	male	False
	743	3rd	45	male	True
	515	2nd	2	male	True
	658	3rd	1	female	True
	837 total ›				
Test	1227	3rd	19	male	False
	188	1st	16	female	True
	397	2nd	34	female	True
	944	3rd	37	female	False
	262	1st	35	male	True
	95	1st	4	male	True
	1080	3rd	22	female	True
	918	3rd	39	male	True
	517	2nd	37	male	False
	425	2nd	30	male	False
	209 total ›				

Figure 8-20. *Titanic dataset divided by train and test set*

Classify Function

The Classify command is another super function used in the Wolfram Language machine learning scheme. This function can be used in tasks that consist of solving a classification problem. The data that this function accepts are numerical, textual, sound, and image data. The input data of this function can be in the same way as with the Predict function {x → y}. However, it is also possible to enter data as a list of elements, as an association of elements, or as a dataset. In this case we will introduce it as a dataset.

In this case we will extract the data from the dataset format by specifying that the columns input (class, age, sex) pointing to the target (survived). Now let's build the classifier function (Figure 8-21), with the following options, Method → {LogisticRegression, L1 → Automatic, L2 → Automatic}. When choosing Automatic, we let Mathematica choose the best combination of L1 and L2 parameters. For the OptimizationMethod set the StochasticGradientDescent method. And for performance goal set Quality. Finally, choosing a seed with a value of 100,000 and the CPU unit as the target device.

The optimization methods for the logistic regression are limited memory Broyden-Fletcher-Goldfarb-Shanno algorithm (LBFGS), StochasticGradientDescent (stochastic gradient method) and Newton (Newton method). These are for estimating the parameters of logistic function. The rule construction will be done from the data inside the dataset using the query language.

```
In[10]:= CF=Classify[Flatten[Values[Normal[Query["Train",All,All,{#class,#a
ge,#sex}→ #survived&][dataset]]]],Method→{"LogisticRegression","L1Regular
ization"→ Automatic,"L2Regularization"→ Automatic,"OptimizationMethod"→"
StochasticGradientDescent"},PerformanceGoal→"Quality",RandomSeeding→10000
0,TargetDevice→"CPU",TrainingProgressReporting→None]
Out[10]=
```

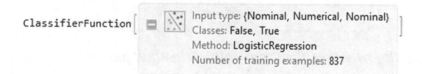

Figure 8-21. *ClassifierFunction object*

After training, like with the Predict function, the Classify function returns a classifier function object (Figure 8-21) instead of a predictor function. Inspecting the classifier function we can see the input data types, which are two—nominal and numerical. The classes, which is the survival status—either false or true. The method used (Logistic Regression); and the number of examples (837). To obtain information on the model use the Information command. Let's look at the model report (Figure 8-22).

```
In[11]:= Information[CF]
Out[11]=
```

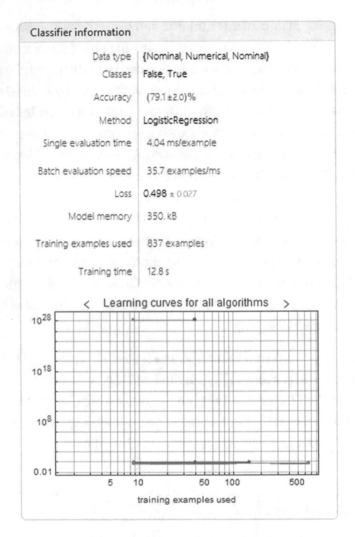

Figure 8-22. *Information about the trained classifier function*

Note   If you click the arrows above the graphs, three plots will be shown: Learning curve, Accuracy, and Learning curve for all algorithm. If you hover the pointer over the line of the last one, a tooltip appears with the corresponding parameters along with the method used, as shown in Figure 8-23 .

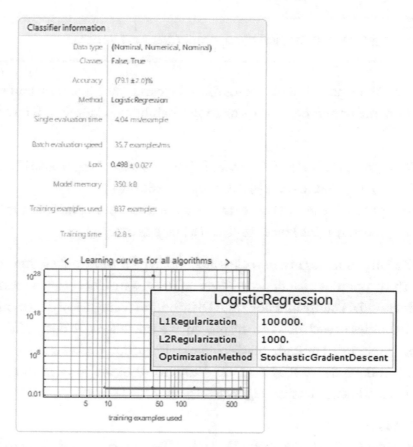

Figure 8-23. *Algorithm specifications tooltip from the method logistic regression*

We see that the model's accuracy is approximately 78%. We also observe by clicking arrows of the plots that the learning curve and accuracy show no signs of improvement from 500 examples and more. To access all the properties of the trained model, add Properties as an option in Information.

```
In[12]:= Information[CF,"Properties"]
Out[12]= {Accuracy,BatchEvaluationSpeed,BatchEvaluationTime,Classes,
ClassNumber,ClassPriors,EvaluationTime,ExampleNumber,FeatureNames,
FeatureNumber,FeatureTypes,FunctionMemory,FunctionProperties,
IndeterminateThreshold,L1Regularization,L2Regularization,LearningCurve,
MaxTrainingMemory,MeanCrossEntropy,Method,MethodDescription,MethodOption,
OptimizationMethod,PerformanceGoal,Properties,TrainingClassPriors,
TrainingTime,UtilityFunction}
```

Note   Depending on the method used, properties may vary.

Let's see what the probabilities are for the data: class = 3rd, age = 23, and sex = male. Probability → name or number of class or TopProbabilities → number of most likely classes.

```
In[13]:= CF[{"3rd",23,"male"},{"Probability"→ False,"TopProbabilities"→ 2}]
Out[13]= {0.839494,{False→0.839494,True→0.160506}}
```
The probabilities of the latter example show that survival status of the passenger may be more inclined to the False status.

To see the full properties of a new classification, type the example followed by Properties. The properties included are Decision (best choice of class according to probabilities and its utility function) and Distribution (categorical distribution object). Probabilities of each class are displayed as associations, ExpectedUtilities (expected probabilities), LogProbabilities (natural logarithm probabilities), Probabilities(probabilities of all classes), and TopProbabilities (most likely class). This is displayed in the following dataset (Figure 8-24).

```
In[14]:= Dataset@
AssociationMap[CF[{"3rd",23,"male"},#] &,{"Decision","Distribution",
"ExpectedUtilities","LogProbabilities","Probabilities","TopProbabilities"}]
Out[14]=
```

Decision	False	
Distribution	CategoricalDistribution[	Input type: Scalar Categories: False True]
ExpectedUtilities	<|False → 0.839494, True → 0.160506, Indeterminate → 0.|>	
LogProbabilities	<|False → -0.174956, True → -1.82943|>	
Probabilities	<|False → 0.839494, True → 0.160506|>	
TopProbabilities	{False → 0.839494, True → 0.160506}	

Figure 8-24. *Properties for the classifier function of the trained model*

Note   To check the logarithm result, use the Log command, Log["base", "number"].

Testing the Model

We will now test the model on the test data using the ClassifierMeasurements (Figure 8-25) command, adding the function and the test set as arguments and the computation of the uncertainty.

```
In[15]:= CM=ClassifierMeasurements[CF,Flatten[Values[Normal[Query["Test",
All,All,{#class,#age,#sex}→ #survived&][dataset]]]],ComputeUncertainty→
True,RandomSeeding→8888]
Out[15]=
```

Figure 8-25. *ClassifierMeasurements object of the classifier function*

The object returned is called a ClassifierMeasurementsObject (Figure 8-25), which is used to look for the properties of the ClassifierFunction after testing the test set. Let's now look at the report (Figure 8-26).

```
In[16]:= CM["Report"]
Out[16]=
```

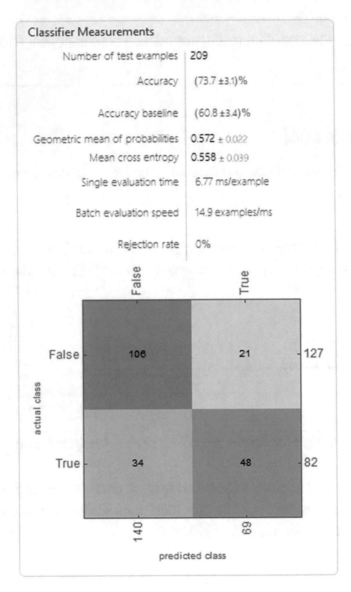

Figure 8-26. *Report of the tested model*

The report seen in the figure shows information such as the number of test examples, the accuracy, and the accuracy baseline, among others. It also shows us the confusion matrix, which shows us the prediction results for the classification model, showing the number of correct and incorrect predictions; these being broken down by class in this case return either false or true, which gives us an idea of the errors the model is making and the type of error it is making. Basically, it shows us the true positives and true negatives and false-positives and false-negatives for each class.

Let's look at the graph (confusion matrix) in a concrete way (Figure 8-27).

```
In[17]:= CM["ConfusionMatrixPlot"]
Out[17]=
```

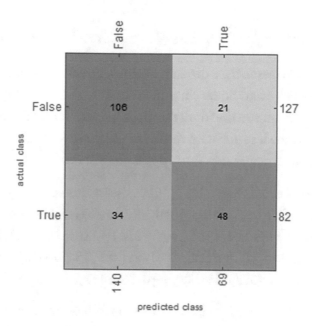

Figure 8-27. *Confusion matrix plot of the tested model*

To get the values of the confusion matrix, use CM["ConfusionMatrix"] or class CM["ConfusionFunction"].

Looking at the plot, we see that the model classified, starting from left to right at the top, 106 examples of false correctly classified, 21 examples of false as true, 34 examples of true as false, and 48 examples of true correctly. To better visualize the performance, let's look at the ROC curves (Figure 8-28) for each class, their respective values, and the Matthews correlation coefficient and AUC values.

```
In[18]:= {CM["ROCCurve"],Dataset@<|{"AUC"→CM["AreaUnderROCCurve"]},
{"MCC"→CM["MatthewsCorrelationCoefficient"]}|>}
```

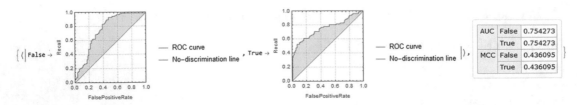

Figure 8-28. *ROC curves for each class along with AUC and MCC values*

Apparently the two classes have the same values, but compared to the ROC curve we can see that the class False had better classification than the True class; let's see the least certain examples so we can see that the True class has worst certain examples than False. With this we can show the less accurate results of the model, which have the highest entropy distribution and mean cross-entropy for each class.

```
In[19]:= CM[{"LeastCertainExamples","ClassMeanCrossEntropy"}]
Out[19]= {{{3rd,39,female}→False,{3rd,38,female}→True,{3rd,37,female}→
False,{3rd,37,female}→False,{3rd,36,female}→True,{3rd,32,female}→False,
{1st,4,male}→True,{3rd,30,female}→False,{3rd,28,female}→False,{3rd,27,
female}→True},<|False→0.363541,True→0.85931|>}
```

To get the values of the MCC coefficient, use the following properties: FalseDiscoveryRate, FalsePositiveRate, FalseNegativeRate (false-positive and false-negative discovery rate for each class), FalseNegativeExamples, FalseNegativeNumber (true negatives), FalsePositiveExamples and FalsePositiveNumber (true positive). These are shown in a short form here.

```
In[20]:= CM[#]&/@{"FalseDiscoveryRate","FalseNegativeRate",
"FalsePositiveRate"}
Out[20]= {<|False→0.242857,True→0.304348|>,<|False→0.165354,
True→0.414634|>,<|False→0.414634,True→0.165354|>}
```

Another way to see if the model behaves consistently in predictions is to look at key metric values like Accuracy, Recall, F1Score Precision, and the Accuracy rejection plot (Figure 8-29). Let's look at these metrics for the model.

```
In[21]:= CM[{"Accuracy","Recall","F1Score","Precision",
"AccuracyRejectionPlot"}]//TableForm
Out[21]//TableForm=
```

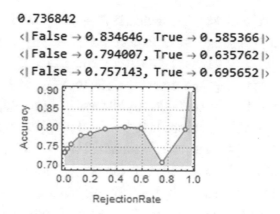

Figure 8-29. *TableForm for the values of Accuracy, Recall, F1Score, Precision, and AccuracyRejectionPlot*

To see related metrics about the accuracy, type the following properties: Accuracy (number of correctly classified examples), AccuracyBaseline (accuracy of predicting the common class), and AccuracyRejectionPlot (ARC plot, accuracy rejection curve). However, to find information about probability and the predicted class of the test set, use the following properties: DecisionUtilities (value of the utility function for every example in the test set), Probabilities (probabilities for every example in the test set), and ProbabilityHistogram (histogram of class probabilities).

Let's see how the probability behaves by plotting the probability of a passenger survival status (Figure 8-30). Remembering that the false state means that a passenger did not survive, and True means that a passenger did survived.

```
In[22]:= TruPlot=
{Plot[{CF[{#1,age,#4},"Probability"]→ #6 ],CF[{#2,age,#4},"Probability"]→
#6 ],CF[{#3,age,#4},"Probability"]→ #6 ]}, {age,0,90},PlotLegends→{"Male
in 1st class", "Male in 2nd class ", "Male in 3rd class"},FrameLabel→
{Style["Age in years",Bold,15], Style["Probability",Bold,15]}, Frame
→#6,FrameTicks→#7,GridLines→ {{20,40,60,80}},ImageSize→#8],Plot[{
CF[{#1,age,#5},"Probability"]→ #6 ],CF[{#2,age,#5},"Probability"]→ #6
],CF[{#3,age,#5},"Probability"]→ #6 ]}, {age,0,90},PlotLegends→{"Female
in 1st class", "Female in 2nd class ", "Female in 3rd class"},FrameLabel→
{Style["Age in years",Bold,15], Style["Probability",Bold,15]}, Frame→#6,
FrameTicks→#7,GridLines→ {{20,40,60,80}},ImageSize→#8]}&["1st","2nd",
"3rd","male","female",True,All,250];
```

```
FalsPlot={Plot[{CF[{#1,age,#4},"Probability"→ #6 ],CF[{#2,age,#4},
"Probability"→ #6],CF[{#3,age,#4},"Probability"→ #6]}, {age,0,90},
PlotLegends→{"Male in 1st class", "Male in 2nd class ", "Male in 3rd
class"},FrameLabel→ {Style["Age in years",Bold,15], Style["Probability",
Bold,15]}, Frame→True,FrameTicks→#7,GridLines→ {{20,40,60,80}},ImageSize
→#8],Plot[{CF[{#1,age,#5},"Probability"→ #6 ],CF[{#2,age,#5},
"Probability"→ #6 ],CF[{#3,age,#5},"Probability"→ #6]},
{age,0,90},PlotLegends→{"Female in 1st class", "Female in 2nd class
", "Female in 3rd class"},FrameLabel→ {Style["Age in years",Bold,15],
Style["Probability",Bold,15]}, Frame→True,FrameTicks→#7,GridLines→
{{20,40,60,80}},ImageSize→#8]}&["1st","2nd","3rd","male","female",False,
All,250];
Headings={Style["True class",Black,20,FontFamily→"Arial Rounded
MT"],Style["False class",Black,20,FontFamily→"Arial Rounded MT"]};
Grid[{{Headings[[1]],Headings[[2]]},{TruPlot[[1]],FalsPlot[[2]]},{TruPlot
[[2]],FalsPlot[[1]]}},Alignment→{{Center,Center},{None,None}},Dividers→
{False,1}]
Out[22]=
```

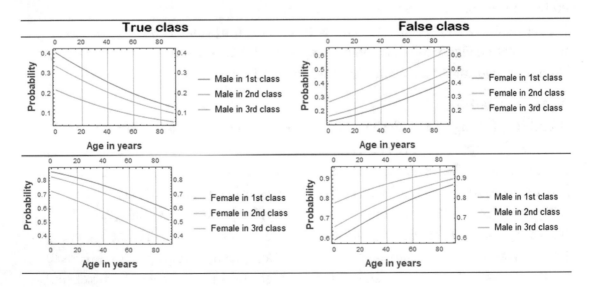

Figure 8-30. *Probabilities of each class, depending on the class, age, and sex*

In the graphs shown in Figure 8-30, clearly it is seen that the probability of survival decreases for males as the age goes up, to even hit values below 20% of chance, whether 1st, 2nd, and 3rd class. This is contrary to the probability of survival for females, where it starts with values above 60% of chance and decreases as age increases too, hitting values above 50% for 1st class.

Data Clustering

The data clustering method is a type of unsupervised learning, as referenced by M. Emre Celebi, and Kemal Aydin in *Unsupervised Learning Algorithms* (2018; Softcover Reprint of the Original 1st 2016 ed. ed.: Springer). It is generally used to find structures and characteristics of data clusters, where the points to be observed are divided into different groups by which they are compared based on unique characteristics.

In the following example, we will create a bivariate data series and plot the list of points (Figure 8-31). To find clusters, there is the Find Clusters command; this command makes a partition of the points according to their similarities.

```
In[1]:= BlockRandom[
SeedRandom[321];
RndPts=Table[{i,RandomReal[{0,1}]},{i,1,450}];]
ListPlot[RndPts,PlotRange→All,PlotStyle→Directive[Thick,Blue],Frame→True,
FrameTicks→All]
Out[1]=
```

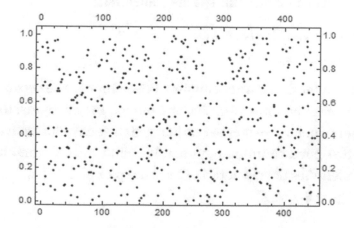

Figure 8-31. *2D scatter plot of random data*

Clusters Identification

The FindClusters function is used to detect partitions within a set of data with similar characteristics. This function gathers the cluster elements into subgroups that the function finds. When you do not add options to the Find Clusters command, the cluster identification parameters will be set automatically by Mathematica. Some options that are used for other machine learning methods can also be used for this command. For example, PerformanceGoal, Method, and RandomSeeding, among others.

```
In[2]:= Clusters=FindClusters[RndPts,PerformanceGoal→"Speed",Method→Autom
atic,DistanceFunction→Automatic,RandomSeeding→1234];
Short[Clusters,4]
Out[2]//Short= {{{1,0.924416},{2,0.695055},{5,0.715785},{8,0.951038},<<137>
>,{372,0.895003},{395,0.917268},{410,0.974659},{422,0.962478}},{<<1>>},{{23
6,<<19>>},<<166>>}}
```

Let's see how many clusters were identified. We will use the Length command; this way we will obtain the general form of the list.

```
In[3]:= Length[Clusters]
Out[3]= 3
```

We see that the result is three. This can be interpreted as follows: the list contains three elements (that is, three sublists), each list represents a cluster, and within each cluster there is a sublist, which contains the points of each identified cluster. To find out how many elements are included in each cluster, we use the Map command, and we apply the Dimension command at the specification level.

```
In[4]:= Map[Dimensions,Clusters,1]
Out[4]= {{145,2},{138,2},{167,2}}
```

This tells us that the first cluster contains 145 elements, the second cluster contains 138 elements, and the third cluster contains 167 elements; these are the same number of points we created earlier, which equal 450. Each cluster is comprised of a two-point coordinate system. The FindClusters command returns the points where it identifies the clusters. Let's see the plot of the clusters generated; this is exhibited in Figure 8-32.

```
In[5]:= ListPlot[Clusters,PlotStyle→{Red,Blue,Green},PlotLegends→
Automatic,Frame→True,FrameTicks→All,PlotLabel→Style["Cluster Plot",Italic,
20,Black],Prolog→{LightYellow,Rectangle[Scaled[{0,0}],Scaled[{1,1}]]}]
Out[5]=
```

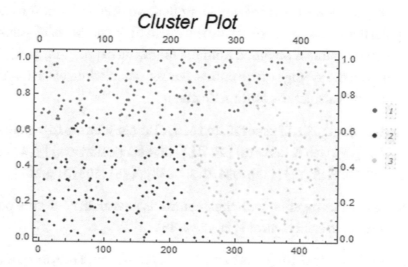

Figure 8-32. *2D scatter plot of the three clusters identified*

As we can see, Find Clusters automatically found the clusters and colored them. To explicitly establish the number of clusters to search, we add the desired number as the second argument—that is, in the form FindCluster ["points", "number of clusters"]. In the previous example we set the method option to automatic. The different methods for finding the clusters are shown here. Agglomerate (which is the algorithm of single linkage clustering), density-based spatial clustering of applications with noise (DBSCAN), NeighborhoodContraction (nearest-neighbor chain algorithm), JarvisPatrick (Jarvis\[Dash]Patrick clustering algorithm), KMeans (k-means clustering), MeanShift (mean-shift clustering), KMedoids (k-medoids partitioning), SpanningTree (minimum spanning tree clustering), Spectral (spectral clustering), and GaussianMixture (Gaussian mixture model).

Choosing a Distance Function

In addition to the method option, there is also the DistanceFunction, which was given the value of Automatic. This option is used to define how the distance between the points is calculated. In general when we choose automatic, the square Euclidean distance is used ($\sum (y_i - x_i)^2$, SquaredEuclideanDistance). There are also other values for the distance function, EuclideanDistance $\left(\sum \sqrt{(y_i - x_i)^2}\right)$, ManhattanDistance ($\sum |x\_\{i\} - y\_\{i\}|$), ChessboardDistance, or ChebyshevDistance ($max(|x\_\{i\} - y\_\{i\}|)$), among others.

Now that we know how the clusters are identified, we want to know the centroid of each one. For this it is necessary to calculate the mean of the points of the clusters. The centroid of a series of points is obtained from the following expression, $\mu = \sum \frac{x_i}{n}$, which can be interpreted as the average of the points. For the calculation, we extract the data from each cluster and calculate its arithmetic mean.

```
In[6]:= {Cluster1Centroid,Cluster2Centroid,Cluster3Centroid}={N@Mean@
Clusters[[1,All]],N@Mean@Clusters[[2,All]],N@Mean@Clusters[[3,All]]}
Out[6]= {{182.807,0.815713},{115.935,0.300888},{353.108,0.39227}}
```

Let´s plot the clusters together with their centroids to visualize how the points are classified with respect to each centroid (Figure 8-33).

```
In[7]:= ClusterPlot=ListPlot[Clusters,PlotStyle→{Red,Blue,Green},
PlotLegends→{"Cluster 1","Cluster 2","Cluster 3"}];
CentroidPlot=ListPlot[{Cluster1Centroid,Cluster2Centroid,Cluster3Centroid},
PlotStyle→Black];
Show[{ClusterPlot,CentroidPlot},Prolog→{LightYellow,Rectangle[Scaled
[{0,0}],Scaled[{1,1}]]},Frame→ True,FrameTicks→ All,PlotLabel→Style
["Cluster Plot",Italic,20,Black]]
Out[7]=
```

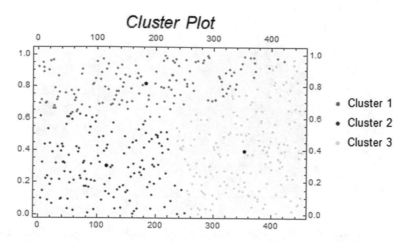

Figure 8-33. *2D scatter plot of the three clusters identified with their respective centroids*

To make sure the first cluster corresponds to the red points, try using ListPlot to plot the points contained in Clusters[[1, All]], as well as those in the second cluster (blue) and third cluster (green).

As an alternative we can highlight the area of the centroids by adding the option Epilog to the plot. Epilog is another graphic option like Prolog, but we will use it to highlight the area of the centroid points (Figure 8-34).

```
In[8]:= Show[{ClusterPlot,CentroidPlot},Prolog→{LightYellow,Rectangle
[Scaled[{0,0}],Scaled[{1,1}]]},Frame→ True,FrameTicks→ All,Epilog→
{Opacity[0.2],PointSize[0.1],Point[Cluster1Centroid],Point
[Cluster2Centroid],Point[Cluster3Centroid]}]
Out[8]=
```

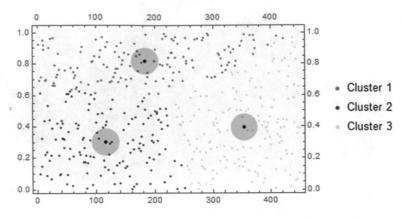

Figure 8-34. *2D scatter plot of the three clusters identified with their respective centroids*

Identifying Classes

Once we have our clusters identified by the command FindClusters, we can use the ClusteringComponents command to label or identify the different classes that were found. We must specify the number of clusters and the specification of where to look for the clusters within the ClusteringComponents command, since there are several ways to use ClusteringComponents.

```
In[9]:= Classes=ClusteringComponents[Clusters,3,2,Method→Automatic,
DistanceFunction→Automatic,RandomSeeding→ 1234,PerformanceGoal→"Speed"]
Out[9]= {{1,1,1,1,1,1,1,1,1,1,1,1,1,1,1,1,1,1,1,1,1,1,1,1,1,1,1,1,1,1,1,1,1,
1,1,1,1,2,1,1,2,1,2,1,2,2,1,1,1,1,2,1,1,2,1,1,2,1,2,2,2,2,1,1,2,2,1,2,2,2,
2,2,2,1,2,2,2,2,1,2,2,2,2,2,2,2,2,2,2,2,2,2,2,2,2,2,2,2,2,2,2,2,2,2,2,2,2,
2,2,2,2,2,2,2,2,2,2,2,2,2,2,2,2,2,2,2,2,2,2,2,2,2,2,2,2,2,2,2,2,2,2,2,2,2,
2,2},{1,1,1,1,1,1,1,1,1,1,1,1,1,1,1,1,1,1,1,1,1,1,1,1,1,1,1,1,1,1,1,1,1,1,
1,1,1,1,1,1,1,1,1,1,1,1,1,1,1,1,1,1,1,1,1,1,1,1,1,1,1,1,1,1,1,1,1,1,1,1,1,
1,1,1,1,1,1,1,1,1,1,1,1,1,1,1,1,1,1,1,1,1,1,1,1,1,3,1,1,3,1,3,1,1,1,1,1,1,
1,3,1,3,3,1,3,1,1,3,1,1,1,3,1,3,1,1,3,3,1,1,3,3,3,3,3,3},{2,3,2,3,3,3,
3,3,3,3,3,3,3,2,3,3,3,3,3,2,3,3,3,3,3,3,2,3,3,3,3,3,3,2,3,3,3,3,3,3,3,
3,3,3,3,3,3,3,3,2,3,3,3,3,3,3,3,3,2,2,2,3,2,2,3,3,2,3,3,3,2,3,3,3,3,3,
2,3,2,3,3,3,2,3,3,2,3,3,2,2,3,3,3,3,3,2,2,2,3,3,3,3,3,3,3,3,3,3,3,3,2,3,
3,3,2,3,3,2,2,3,2,3,3,3,3,3,2,3,3,3,3,3,2,3,3,3,2,3,3,3,2,3,2,2,3,2,3,3,2,
2,3,3,2,2,2,3,3,3,3,2,3,3}}
```

In this way, numbers that correspond to the three classes appear. The command only identifies that there are three types of classes; it does not mention what each class means. This is because cluster methods are often performed on unlabeled data, so interpretation is performed as part of the analysis. Let's count how many elements of each class we have.

```
In[10]:= Flatten[Classes]//Counts
Out[10]= <|1→174,2→132,3→144|>
```

The command returns us that class one contains 174, class two contains 132, class three contains 144. One point to clarify is that why the clusters identified with FindClusters and ClusteringCompnents defer. Well, this is because by setting the automatic option in the distance function, we are telling Mathematica to find the optimal distance function. And depending on the data one function might gather elements in different forms as we will see later on.

K-Means Clustering

At the moment we have seen how to search for clusters in a generic way. In this part we will focus on the K-means method.

The K-means is a technique to find and classify groups (k) of data so that the elements that share similar characteristics are grouped together and in the same way for the opposite case (not similar characteristics). To distinguish whether the data contain similarities or not, the method calculates the distance between the data with respect to a centroid. The elements that have less distance between them will be those that share similarities. This technique is carried out as an iterative process in which the groups are adjusted until they reach a convergence. Basically, the K-means method, which is a simple algorithm, consists of making a classification by means of specific partitions, in different groups, where each point or observation belongs to the group. Clustering is done by minimizing the sum of the distances between each object and the centroid of its group. The k-means clustering technique tries to build the clusters so that they have the least variation within a group. This is done by minimizing the expression

$J(C_i) = \sum_{x_j \in C_i}^{N} \|x_j - \mu_i\|^2$, where C_i represents the i-th cluster, x_j represents the points,

and μ_i represents the centroid of each cluster C_i. The square term of the function is

the distance function; the most used is the square Euclidean distance, as in this case.

To learn more about the mathematical foundation behind this technique, consult the reference *An Introduction to Statistical Learning: With Applications in R* by Gareth James, Daniela Witten, Trevor Hastie, and Robert Tibshirani. (1st ed. 2013, Corr. 7th printing 2017 ed.: Springer).

In the following example we will use the Fisher's Irises dataset found in ExampleData. Remembering the features that this dataset has, execute the following code.

```
In[11]:= ExampleData[{"Statistics","FisherIris"},"ColumnDescriptions"]
Out[11]= {Sepal length in cm.,Sepal width in cm.,Petal length in cm.,Petal
width in cm.,Species of iris}
```

Let's extract the dataset and assign the variable iris to it.

```
In[12]:= iris=ExampleData[{"Statistics","FisherIris"}];
Take a look at the dataset.
```

```
In[13]:= Short[iris,6]
Out[13]//Short= {{5.1,3.5,1.4,0.2,setosa},{4.9,3.,1.4,0.2,setosa},{4.7,3.2
,1.3,0.2,setosa},{4.6,3.1,1.5,0.2,setosa},{5.,3.6,1.4,0.2,setosa},{5.4,3.9
,1.7,0.4,setosa},{4.6,3.4,1.4,0.3,setosa},{5.,3.4,1.5,0.2,setosa},{4.4,2.9
,1.4,0.2,setosa},{4.9,3.1,1.5,0.1,setosa},{5.4,3.7,1.5,0.2,setosa},{4.8,3.
4,1.6,0.2,setosa},<<126>>,{6.,3.,4.8,1.8,virginica},{6.9,3.1,5.4,2.1,virgi
nica},{6.7,3.1,5.6,2.4,virginica},{6.9,3.1,5.1,2.3,virginica},{5.8,2.7,5.1
,1.9,virginica},{6.8,3.2,5.9,2.3,virginica},{6.7,3.3,5.7,2.5,virginica},{6
.7,3.,5.2,2.3,virginica},{6.3,2.5,5.,1.9,virginica},{6.5,3.,5.2,2.,virgini
ca},{6.2,3.4,5.4,2.3,virginica},{5.9,3.,5.1,1.8,virginica}}
```

Dimensionality Reduction

Since the iris dataset consists of four features that are classified into three types of species, we will use the PCA method, as this method is used to reduce high-dimensionality problems. In this case, what we want is to represent these features through two main components. For this we proceed to standardize the data—that is, they have zero mean and standard deviation 1, since the variables with larger variance are more likely to affect the PCA.

```
In[14]:= ST=Standardize[iris[[All,{1,2,3,4}]]];(*Showing only the first 4
terms*)
%[[1;;4]]//TableForm
Out[14]//TableForm=
```

-0.897674	1.0156	-1.33575	-1.31105
-1.1392	-0.131539	-1.33575	-1.31105
-1.38073	0.327318	-1.3924	-1.31105
-1.50149	0.0978893	-1.2791	-1.31105

There are two ways to do the process, either using the DimensionReduce command
or the DimensionReduction command, which are used to reduce the dimensions of
the data. The difference between the two is that the first returns the values as a list.
The second returns a DimensionReducerFunction (Figure 8-35) as output as in the
case of Predict and Classify. Both belong to the Wolfram Language special functions
for machine learning. For this case we will use the DimensionReduction command.
Since we have the data, we introduce the standardized data as arguments, followed by
specified target dimensions (2), with the as "PrincipalComponentAnalysis" method. This
will give us the DimensionReducerFunction that will assign us the name of DR.

```
In[15]:= DR=DimensionReduction[ST,2,Method→"PrincipalComponentsAnalysis"]
Out[15]=
```

Figure 8-35. *DimensionReductionFunction object*

The properties of the function are "ReducedVectors" (list of reduced vectors),
"OriginalData" (deduction from the original data list given the reduced vectors),
"ReconstructedData" (data reconstruction by reduction and inversion), "ImputedData"
(missing values replaced by imputed ones). We call the function for the standardized
data values, showing the first five. The coordinates x and y will be for the principal
components 1 and 2, respectively.

```
In[16]:= PCA=DR[ST,"ReducedVectors"];
TableForm[%[[1;;5]],TableHeadings→{None, {"First Principal
Component","Second Principal Component"}},TableAlignments→Center]
Out[16]//TableForm=
```

First Principal Component	Second Principal Component
-2.2647	0.480027
-2.08096	-0.674134
-2.36423	-0.341908
-2.29938	-0.597395
-2.38984	0.646835

This calculates the variance of each component, followed by the total to find the proportion of variance explained. Observing that PC1 seems to represent 76% of the data dispersion, and PC2 seems to represent 23%. To obtain the accumulated percentage we add the variations of each component. To view more depth about the proportion of variation refer to *An Introduction to Statistical Learning: With Applications in R* (James, G., Witten, D., Hastie, T., & Tibshirani, R. ; 1st ed. 2013, Corr. 7th printing 2017 ed.: Springer).

```
In[17]:=
Variance@PCA[[All,All]]/Total@Variance@PCA[[All,All]]//TableForm[#,TableHea
dings→{{"First PC variation","Second PC variation"}, None}]&
Out[17]//TableForm=
```

First PC variation	0.761507
Second PC variation	0.238493

We look at the plot (Figure 8-36) of the main components made by the previous process. If you look over the complete iris data from the ExampleData, the first 50 elements correspond to the setosa specie, the next 50 to versicolor, and the last 50 to virginica.

```
In[18]:= Labels={Style["First principal component",Black,Bold],Style["Seco
nd Principal component",Black,Bold]};
ListPlot[{PCA[[1;;50]],PCA[[51;;100]],PCA[[100;;150]]},PlotLegends→Placed[
{Placeholder["setosa"],Placeholder["versicolor"],Placeholder["virginica"]},-
Right],PlotMarkers→"OpenMarkers",GridLines→All,Frame→True,Axes→False,Fr
ameTicks→All,FrameLabel→Labels]
Out[18]=
```

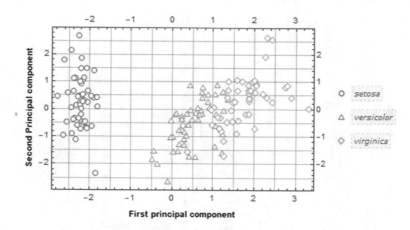

Figure 8-36. *Scatter plot of the two principal components*

Applying K-Means

Now let's find the clusters with K-means, using the Manhattan distance. By specifying to look for three clusters; we are making the assumption that the data can be divided into three clusters. This is because we know that the original data belongs to three species (setosa, versicolor, and virginica). The plot of the clusters is shown here (Figure 8-37), with their respective centroids. When choosing the k-means method, suboptions can be added, like InitialCentroids. Costume start centroids (a list of centroid coordinates) can be typed or we can leave the automatic option. To enter the centroids coordinates, we use the following form Method → {"KMeans","InitialCentroids" → {{x1,y1}, {x2,y2}, {x3,y3} ... }}, where x1, y1 represent the centroid of the C1 (cluster 1). Initial centroids will not be given to the command FindClusters to keep some sort of randomness.

```
In[19]:= Clstr=FindClusters[PCA,3,Method→"KMeans",DistanceFunction→Square
dEuclideanDistance,RandomSeeding→8888];
ListPlot[Clstr,PlotRange→All,Frame→True,AspectRatio→0.8,Axes→False,Pl
otStyle→{ColorData[97,1],ColorData[97,2],ColorData[97,3]},PlotLabel→Sty
le["K-means clustering for K=3",FontFamily→"Times",Black,20,Italic],Fra
meTicks→All,PlotLegends→Placed[{Placeholder[Style["Cluster 1",Bold,Bla
ck,10]],Placeholder[Style["Cluster 2",Bold,Black,10]],Placeholder[Style[
"Cluster 3",Bold,Black,10]]},Right],PlotMarkers→ "OpenMarkers",FrameLab
el→Labels,GridLines→All,Epilog→{Opacity[1],PointSize[0.01],Point[Mean@
Clstr[[1,All]]],Point[Mean@Clstr[[2,All]]],Point[Mean@Clstr[[3,All]]]}]
Out[19]=
```

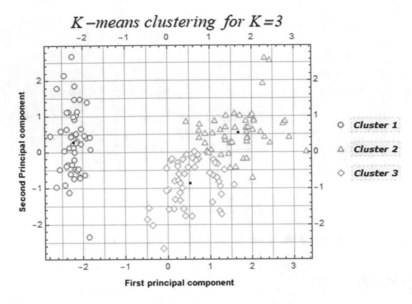

Figure 8-37. *3 clusters identified of the two principal components*

In Figure 8-37, it appears that the method clearly identifies the left points as a single cluster (setosa specie), whereas some of the points between clusters 2 and 3 might be misclassified.

Chaining the Distance Function

Changing the DistanceFunction can modify how the clusters are arranged, the next code shows the plot for k = 3 and choosing a different distance function. In the next block of code, the computation of the clusters is made for the same k (3), with a different distant function and stored into their respective variables. Then the clusters are plotted (Figure 8-38) for each of the different distance functions, and finally they are displayed within a graphic grid.

```
In[20]:= {ED,MhD,ChD,CosD}={FindClusters[PCA,3,PerformanceGoal→#1,Method→
#2,DistanceFunction→EuclideanDistance,RandomSeeding→#3],FindClusters[PCA,
3,PerformanceGoal→#1,Method→#2,DistanceFunction→ManhattanDistance,Random
Seeding→#3],FindClusters[PCA,3,PerformanceGoal→#1,Method→#2,DistanceFunc
tion→ChessboardDistance,RandomSeeding→#3],FindClusters[PCA,3,Performance
Goal→#1,Method→#2,DistanceFunction→CosineDistance,RandomSeeding→#3]}&
["Quality","KMeans",8888];
{EDplt,MhDplt,ChDplt,CosDplt}={
```

```
ListPlot[ED,Frame→#1,AspectRatio→#2,PlotMarkers→#3,PlotStyle→#4,GridLin
es→#5,PlotRange→#6,ImageSize→#7,FrameLabel→#8,Axes→#9,FrameTicks→#10,
Epilog→{Opacity@#11,PointSize@#12,Point[Mean@ED[[1,All]]],Point[Mean@
ED[[2,All]]],Point[Mean@ED[[3,All]]]},PlotLabel→ Style["Euclidean
Distance",Black]],
ListPlot[MhD,Frame→#1,AspectRatio→#2,PlotMarkers→#3,PlotStyle→#4,GridLi
nes→#5,PlotRange→#6,ImageSize→#7,FrameLabel→#8,Axes→#9,FrameTicks→#10,
Epilog→{Opacity@#11,PointSize@#12,Point[Mean@MhD[[1,All]]],Point[Mean@
MhD[[2,All]]],Point[Mean@MhD[[3,All]]]},PlotLabel→ Style["Manhattan
Distance",Black]],
ListPlot[ChD,Frame→#1,AspectRatio→#2,PlotMarkers→#3,PlotStyle→#4,GridLi
nes→#5,PlotRange→#6,ImageSize→#7,FrameLabel→#8,Axes→#9,FrameTicks→#10,
Epilog→{Opacity@#11,PointSize@#12,Point[Mean@ChD[[1,All]]],Point[Mean@
ChD[[2,All]]],Point[Mean@ChD[[3,All]]]},PlotLabel→ Style["Chessborad
Distance",Black]],
ListPlot[CosD,Frame→#1,AspectRatio→#2,PlotMarkers→#3,PlotStyle→#4,
GridLines→#5,PlotRange→#6,ImageSize→#7,FrameLabel→#8,Axes→#9,Frame
Ticks→#10,
Epilog→{Opacity@#11,PointSize@#12,Point[Mean@CosD[[1,All]]],Point[Mean@
CosD[[2,All]]],Point[Mean@CosD[[3,All]]]},PlotLabel→ Style["Cosine
Distance",Black]]
}&[True,0.8,"OpenMarkers",{ColorData[97,1],ColorData[97,2],ColorData[97,3]}
,All,Automatic,300,Labels,False,All,1,0.03];
LegendsText={Placeholder[Style["Cluster 1",Bold,Black,10]],Placeho
lder[Style["Cluster 2",Bold,Black,10]],Placeholder[Style["Cluster
3",Bold,Black,10]]};
Labeled[Legended[GraphicsGrid[{{EDplt,MhDplt},{ChDplt,CosDplt}},Frame→All,
Background→White,Spacings→1],PointLegend[{ColorData[97,1],ColorData[97,2]
,ColorData[97,3]},LegendsText,LegendMarkers→"OpenMarkers"]],Style["K-means
clustering for K=3",FontFamily→"Times",Black,20,Italic],Top]
Out[20]=
```

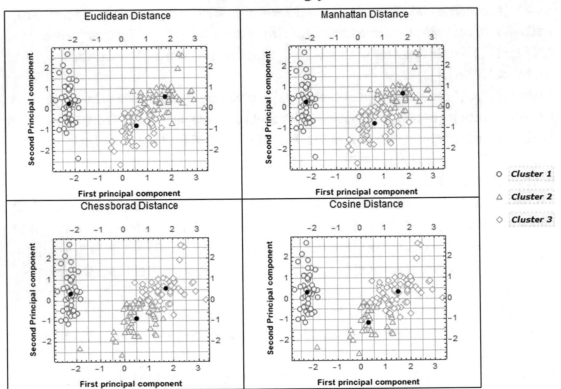

Figure 8-38. *K-means clustering for K = 3, for different distance functions*

As seen in Figure 8-38, the clusters can have different arrangements with different distance functions; one thing to note also is that the clusters centroids change in each of the subfigures (Figure 8-38).

Different K's

Having seen that for different distance functions the clusters can vary, let's now construct the process but with different K's—that is, for k= 2, 3, 4, and 5, as exhibited in Figure 8-39.

```
In[21]:= {K2,K3,K4,K5}={FindClusters[PCA,2,PerformanceGoal→#1,Method→#2,
DistanceFunction→#3,RandomSeeding→#4],FindClusters[PCA,3,PerformanceGoal
→#1,Method→#2,DistanceFunction→#3,RandomSeeding→#4],FindClusters[PCA,4,
PerformanceGoal→#1,Method→#2,DistanceFunction→#3,RandomSeeding→#4],
```

```
FindClusters[PCA,5,PerformanceGoal→#1,Method→#2,DistanceFunction→#3,
RandomSeeding→#4]}&["Speed","KMeans",SquaredEuclideanDistance,8888];
{PK2,PK3,PK4,PK5}={
ListPlot[K2,Frame→#1,AspectRatio→#2,PlotMarkers→#3,PlotStyle→#4,GridLin
es→#5,PlotRange→#6,ImageSize→#7,FrameLabel→#8,Axes→#9,FrameTicks→#10,
Epilog→{Opacity@#11,PointSize@#12,Point[Mean@K2[[1,All]]],Point[Mean@
K2[[2,All]]]},PlotLabel→ Style["K=2",Black]],
ListPlot[K3,Frame→#1,AspectRatio→#2,PlotMarkers→#3,PlotStyle→#4,GridLin
es→#5,PlotRange→#6,ImageSize→#7,FrameLabel→#8,Axes→#9,FrameTicks→#10,
Epilog→{Opacity@#11,PointSize@#12,Point[Mean@K3[[1,All]]],Point[Mean@
K3[[2,All]]],Point[Mean@K3[[3,All]]]},PlotLabel→ Style["K=3",Black]],
ListPlot[K4,Frame→#1,AspectRatio→#2,PlotMarkers→#3,PlotStyle→#4,GridLin
es→#5,PlotRange→#6,ImageSize→#7,FrameLabel→#8,Axes→#9,FrameTicks→#10,
Epilog→{Opacity@#11,PointSize@#12,Point[Mean@K4[[1,All]]],Point[Mean@
K4[[2,All]]],Point[Mean@K4[[3,All]]],Point[Mean@K4[[4,All]]]},PlotLabel→
Style["K=4",Black]],
ListPlot[K5,Frame→#1,AspectRatio→#2,PlotMarkers→#3,PlotStyle→#4,GridLin
es→#5,PlotRange→#6,ImageSize→#7,FrameLabel→#8,Axes→#9,FrameTicks→#10,
Epilog→{Opacity@#11,PointSize@#12,Point[Mean@K5[[1,All]]],Point[Mean@
K5[[2,All]]],Point[Mean@K5[[3,All]]],Point[Mean@K5[[4,All]]],Point[Mean@
K5[[5,All]]]},PlotLabel→ Style["K=5",Black]]
}&[True,0.8,"OpenMarkers",{ColorData[97,1],ColorData[97,2],ColorData[97,3],
ColorData[97,4],ColorData[97,5]},All,Automatic,260,Labels,False,All,1,
0.015];
LegendsText2={Placeholder[Style["Cluster 1",Bold,Black,10]],Placeholder
[Style["Cluster 2",Bold,Black,10]],Placeholder[Style["Cluster 3",Bold,
Black,10]],Placeholder[Style["Cluster 4",Bold,Black,10]],Placeholder[Style
["Cluster 5",Bold,Black,10]]};
Labeled[Legended[GraphicsGrid[{{PK2,PK3},{PK4,PK5}},Frame→All,Background→
White,Spacings→1],PointLegend[{ColorData[97,1],ColorData[97,2],ColorData
[97,3],ColorData[97,4],ColorData[97,5]},LegendsText2,LegendMarkers→
"OpenMarkers"]],Style["K-means clustering for K=2,3,4,5",FontFamily→
"Times",Black,20,Italic],Top]
Out[21]=
```

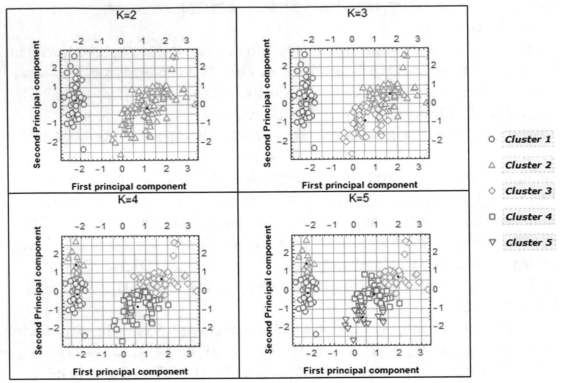

Figure 8-39. *K-means for K from 2 to 5*

As seen in the Figure 8-39, the arrangement of the clusters also depends on the number of k's. Complementing with ClusteringComponents, we can count the number of labels register for a k = 3.

```
In[22]:=ClusteringComponents[Clstr,3,2,Method→"KMeans",DistanceFunction→S
quaredEuclideanDistance,RandomSeeding→8888]
Out[22]={{1,1,1,1,1,1,1,1,2,1,1,1,1,1,1,1,1,1,1,1,1,1,1,1,1,2,1,1,1,1,1,1,
1,1,1,1,1,1,1,1,1,2,1,1,1,2,1,1,1,1},{3,3,3,3,3,3,3,3,3,3,3,3,3,3,3,3,3,3,3
,3,3,3,3,3,3,3,3,3,3,3,3,3,3,3,3,3,3,3,3,3,3,3,3,3,3,3,3,3,3,3},{2,3,2,
2,2,2,3,2,3,2,3,2,3,2,3,3,3,3,3,2,2,2,2,3,2,3,2,2,2,3,2,2,2,2,2,3,2,2,3,2,3
,3,3,3,3,3,3,3,3}}
```

```
In[23]:= Counts[Flatten[%]]
Out[23]= <|1→46,2→29,3→75|>
```

Essentially, given a clustering problem, k-means technique is meant to be used for unlabeled data—that is, data without defined categories. Some factors that can alter the operation of the method include the following.

- The spread, or how far apart the points are. This is reflected if the data contains outliers, which can be erroneously classified as part of a cluster, when visually the opposite is observed.

- The dimensionality of the data. Given that more information and features are often added to the model, the number of dimensions grows. This type of problem can be solved using data transformation methods, as in the example seen from PCA, but with some restrictions, since the PCA method can have a loss of sensitive information on the features.

- The value of k is determined manually, but when there are high values of the cost function, it can be interpreted that the intervariation of the clusters is high, and with low values of the cost function the intervariation of the clusters is low. The last two assumptions can also be attributed to the fact that for lower values of k, many observations can be grouped into large individual clusters, and for high values of k observations they can be a proper group.

Cluster Classify

Another command that belongs to the cluster functions is called ClusterClassify (Figure 8-40). This command works in the same way as Classify does. In the next example we will use this command to see how the k-means cluster classifies the species based on two features: Sepal length and Sepal width. We will split the data into halves when we randomly sample.

```
In[24]:=
 BlockRandom[
SeedRandom[88888];
RandomSample[iris[[All,{1,2}]]];
]
TrS=%[[1;;75]];
TsT=%%[[76;;150]];
```

```
In[25]:= CC=ClusterClassify[TrS,3,Method→"KMeans",DistanceFunction→
Automatic,PerformanceGoal→"Speed",RandomSeeding→8888 ]
Out[25]=
```

```
ClassifierFunction[ Input type: NumericalVector (length: 2)
                    Classes: 1, 2, 3
                    Method: KMeans
                    Number of training examples: 75 ]
```

Figure 8-40. *ClassifierFunction of the cluster classification model*

Getting the classifier function (Figure 8-40), we can see the details of the classifier, and we can see the input vector is a numerical vector, the number of classes (three), the method, and the number of training examples.

Note To correctly use the -means method, the number of clusters needs to be specified; otherwise the command will not execute correctly.

To see information about the classifier function, use Information (Figure 8-41).

```
In[26]:= Information[CC]
Out[26]=
```

Classifier information

Data type	NumericalVector (length: 2)
Classes	1, 2, 3
Method	KMeans
Single evaluation time	3.58 ms/example
Batch evaluation speed	73.5 examples/ms
Model memory	30.3 kB
Training examples used	75 examples
Training time	5.97 ms

Figure 8-41. *Classifier information for K-means*

More detailed information about the classifier function is shown in Figure 8-41. To get the full list of properties, type "Properties"as a second argument. Many metrics, such as BatchEvaluationSpeed, BatchEvaluationTime, and TrainingTime, can be used to compare times with different methods.

```
In[27]:= Information[CC,"Properties"]
Out[27]= {BatchEvaluationSpeed,BatchEvaluationTime,Classes,ClassNumber,
ClassPriors,DistanceFunction,EvaluationTime,ExampleNumber,FeatureNames,
FeatureNumber,FeatureTypes,FunctionMemory,FunctionProperties,Indeterminate
Threshold,LearningCurve,MaxTrainingMemory,Method,MethodDescription,
MethodOption,PerformanceGoal,Properties,TrainingClassPriors,TrainingTime,
UtilityFunction}
```

Let's now get the information about the classes identified from the cluster classifier, the number of classes, distance function, feature names, and the training class probabilities.

```
In[28]:= Information[CC,#]&/@{"Classes","ClassNumber","DistanceFunction",
"FeatureNames","TrainingClassPriors"}
Out[28]= {{1,2,3},3,EuclideanDistance,
{f1},<|1→0.373333,2→0.293333,3→0.333333|>}
```

We can see that there are three classes: class 1, class 2 and class 3. The distance function used is EuclideanDistance, and the numeric vector features are referred to by the name f1. A simple example is used, by choosing a sepal length of 1 and sepal width of 2, to show the different properties that can be used when testing the data; this is shown in the dataset form (Figure 8-42). The example is first written followed by the properties Decision (cluster that belongs the example), Distribution (categorical distribution object for histogram plots), ExpectedUtilities (expected probabilities and indeterminate threshold), LogProbabilities (log probabilities), Probabilities (probabilities of the test data based on classes), and TopProbabilities (best probabilities for the test data).

```
In[29]:= Dataset[AssociationMap[CC[{1,2},#]&,{"Decision","Distribution",
"ExpectedUtilities","LogProbabilities","Probabilities","TopProbabilities"}]]
Out[29]=
```

Decision	1
Distribution	CategoricalDistribution[Input type: Scalar Categories: 1 2 3]
ExpectedUtilities	⟨| 1 → 0.976148, 2 → 2.72758 × 10⁻¹⁶, 3 → 0.0238517, Indeterminate → 0.|⟩
LogProbabilities	⟨| 1 → -0.0241407, 2 → -35.8379, 3 → -3.7359|⟩
Probabilities	⟨| 1 → 0.976148, 2 → 2.72758 × 10⁻¹⁶, 3 → 0.0238517|⟩
TopProbabilities	{1 → 0.976148}

Figure 8-42. *Dataset of the simple example*

We can see that the example belongs to the third cluster and that the associated probability is 1 → 0.976148. Let's look at the rest of the data and plot the cluster classification.

The classified data plot is shown in Figure 8-43.

```
In[30]:= ListPlot[Pick[TsT,CC[TsT],#]&/@{1,2,3},PlotMarkers→"OpenMarkers",
GridLines→Automatic,PlotLegends→{Placeholder[Style["Cluster 1",Bold,
Black,10]],Placeholder[Style["Cluster 2",Bold,Black,10]],Placeholder
[Style["Cluster 3",Bold,Black,10]]},Frame→True,FrameTicks→All,
FrameLabel→{"Sepal Lenght","Sepal Width"}]
Out[30]=
```

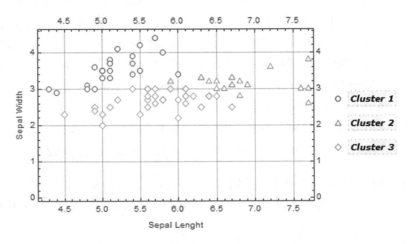

Figure 8-43. *Cluster classification on the example of the iris data for the first two features*

As a complement, a probability restriction for values below an established probability value can be added, with IndeterminateThreshold, as depicted in Figure 8-44.

```
In[31]:= ListPlot[Pick[TsT,CC[TsT,IndeterminateThreshold→ 0.6],#]&/@{1,2,
3,Indeterminate},PlotMarkers→ "OpenMarkers",PlotLegends→{Placeholder
[Style["Cluster 1",Bold,Black,10]],Placeholder[Style["Cluster 2",Bold,
Black,10]],Placeholder[Style["Cluster 3",Bold,Black,10]],Placeholder
[Style["Indeterminate",Bold,Black,10]]},Frame→True,FrameTicks→All,
FrameLabel→{"Sepal Lenght","Sepal Width"},GridLines→Automatic]
Out[31]=
```

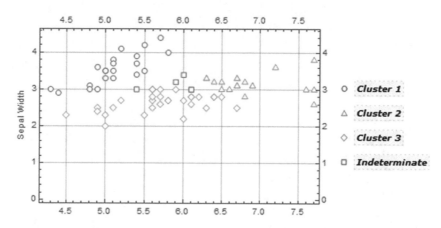

Figure 8-44. *Cluster classification on the example of the iris data for the first two features with a probability restriction*

CHAPTER 9

Neural Networks with the Wolfram Language

In this block will start with the basic foundations of the neural network framework in the Wolfram Language. The chapter starts with the concepts of layers, how to use the commands for different layers, and the most common layers. We will learn how to enter data into the layers by the net port, as well as the different forms of equivalent expression of the layers. This is followed by how to distinguish different layers by their symbol. We will see that layers can have multiple options that enable the layer to have various specifications by viewing the concept of a layer in the Wolfram Language scheme, comparing different layers that have different purposes and that perform different computations. We will also achieve this by looking at the various activation functions that are supported by the Wolfram Language and inspecting the plots of each of the functions in addition to different syntax forms. Next we will view the concepts of encoders and decoders and how these tools are used for the construction of a neural network model, depending on the task to fulfill. We then learn how these encoders and decoders are used to convert different data types to numeric arrays, as well as how to convert the numeric arrays back to the initial data. We introduce the concept of a container and what it means for the created models and what types exist. We will see how to handle and build containers with different commands and how to graphically visualize the created model. We show how the Wolfram Neural Net Framework supports MXNet related operations and how to export a network to the format of the MXNet operation.

Layers

To build a neural network in the Wolfram Language, it is necessary to understand that these are built from layers. A layer is a term that can be applied to a collection of nodes that operate together at a specific level within the neural network. The layer is an essential and simple member that exists for the construction of a neural network.

Input Data

The data handled by the layers is of a numeric type, and not of another type. Input variables can be: vector, a unidimensional list; matrices, a two-dimensional list; and arrays, a list of lists or any other numeric tensor. These input variables can be either features or attributes of the dataset of study, with a known shape, or a multidimensional shape. These types of input attributes are associated with the input layer, for which the feature size, in turn, must be equal to input size of a layer, but not every layer receives the same input and returns the same output; every input varies depending on the type of layer to be used. This definition is one of the most basic idea in neural networks since they are a crucial component of the whole structure that involves the term neural network. A remark here is do not confuse input with input layer since they do not mean the same.

Linear Layer

A linear layer is the most common and widely used layer in a neural network. To build the simplest layer in the Wolfram Language use the command LinearLayer.

```
In[1]:= LinearLayer["Input"→ 1,"Output"→ 2]
Out[1]=
```

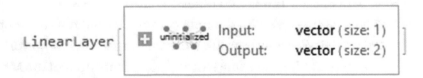

Figure 9-1. *LinearLayer object*

Figure 9-1 represents the LinearLayer object in the Wolfram Language. Clicking on the plus icon shows the internal parameters, including details about the layer port's input and output and array rank of the weights and biases of the linear layer. This is shown in Figure 9-2.

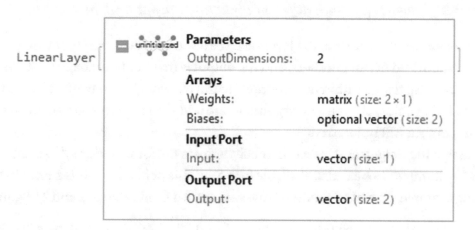

Figure 9-2. *Expanded LinearLayer object*

Each layer has an input port and an output port. Each port has an associated size of what is entering the layer and what is going out. In the latter case, a vector of size one is entering, and the layer returns a vector of size two.

Weights and Biases

The general form of a linear layer is given by the following expression of the dot product $\mathbf{w} \cdot \mathbf{x} + \mathbf{b}$, where $\mathbf{x}$ is the vector of the data, $\mathbf{w}$ represents the matrix of the weights, and $\mathbf{b}$ the vector of the biases. Linear layers have other associated names like fully connected layer, as in the MXnet framework. The input of the layers in the Wolfram Language receives numerical tensors as input—that is, they only act on numerical arrays.

To explicitly enter the size of input and output, we write the form of the input port and the output port followed by different options: "Input" or "Output" → {size, Options.}. Options include defining a real number (Real), a vector of form n (single number n), an array ({n1 * n2 * n3} ...), or a NetEncoder, which we will see later. Following is some equivalent ways to write layers, as depicted in Figure 9-3.

```
In[2]:= LinearLayer["Input"->{2,"Real"},"Output"->{2,1}]
Out[2]=
```

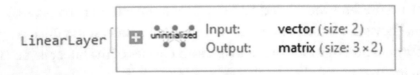

Figure 9-3. *LinearLayer with different input and output rank arrays*

As we can see in Figure 9-3, the layer receives a vector of size two (list of length 2), which is comprised of real numbers, and the output is matrix of the shape 3 x 2. When a real number is specified within the Wolfram Neural Network Framework, it works with the precision of a Real32. When no arguments are added to the layer, the shape of the input and output will be inferred.

To assign the weights and biases manually, write in the form "Weights" → number, "Biases" → number; None is also available for no weights or biases. This is shown in the following example, where weights and biases are set to a fixed value of 1 and 2 (Figure 9-4).

```
In[3]:= LinearLayer["Input"→ 1,"Output"→ 1,"Weights"→ 1,"Biases"→ 2]
Out[3]=
```

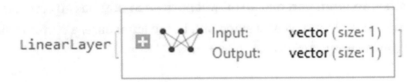

Figure 9-4. *Initialized linear layer, with fixed biases and weights*

Initializing a Layer

Besides being able to initialize the layer ourselves, there is another command that allows us to initialize the layer with random values; this is NetInitialize. So, to establish hold values of weights or biases, we can also use the LearningRateMultipliers option (Figure 9-5). Besides this, LearningRateMultipliers also marks the rate at which a layer learns during the training phase.

```
In[4]:= NetInitialize[LinearLayer["Input"→ "Real","Output"→ "Real",Learni
ngRateMultipliers→{"Biases"→1}]]
Out[4]=
```

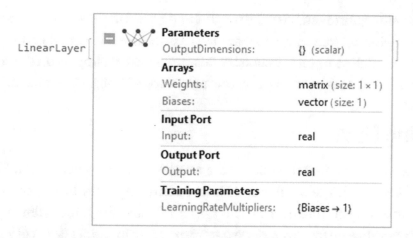

Figure 9-5. *LinearLayer with training parameters*

When a layer is initialized, the uninitialized text disappears. If we observe the properties of the new layer, it will appear within the training parameters that fixed biases have been established and a rate of learning.

The options for NetInitialize are Method and RandomSeeding. The available methods are Kaiming, Xavier, Orthogonal (orthogonal weights), and Random (weights selection from a distribution). For example, we can use the Xavier initialization sampling from a normal distribution, as can be seen in Figure 9-6.

```
In[5]:= NetInitialize[LinearLayer["Input"→ "Real","Output"→ "Real",Learni
ngRateMultipliers→{"Biases"→1}],Method→ {"Xavier","Distribution"→"Norma
l"},RandomSeeding→888]
Out[5]=
```

Figure 9-6. *LinearLayer initialized with the Xavier method*

Note   To see the options set for a layer, the Option command is recommended.

Despite being able to establish the weights and biases manually, it is advisable to start the layer with random values to maintain a certain level of complexity in the overall structure of a model, since on the contrary this could have an impact on the creation of a neural network that does not make accurate predictions for non-linear behavior.

Retrieving Data

NetExtract is used to retrieve the value of the weights and biases in the form NetExtract [net, {level1, level2, ...}]. The weights and biases parameters of the linear layers are packed in NumericArray objects (Figure 9-7). This object will have the values, dimensions, and type of the values in the layer. NetExtract also serves to extract layers of a network with many layers. NumericArrays are used in the Wolfram Language to reduce memory consumption and computation time.

```
In[6]:= LinearL=NetInitialize[LinearLayer[2, "Input"→ 1],RandomSeeding→888];
NetExtract[LinearL,#]&/@{"Weights","Biases"}//TableForm
Out[6]=
```

$$\text{NumericArray}\left[\ \boxplus\ \begin{array}{l}\text{Type: Real32}\\ \text{Dimensions: \{2, 1\}}\end{array}\ \right]$$

$$\text{NumericArray}\left[\ \boxplus\ \begin{array}{l}\text{Type: Real32}\\ \text{Dimensions: \{2\}}\end{array}\ \right]$$

Figure 9-7. *Weights and biases of a linear layer. With Normal we convert them to lists.*

```
In[7]:= TableForm[SetPrecision[{{Normal[NetExtract[LinearL,"Weights"]]
},{Normal[NetExtract[LinearL,"Biases"]]}},3],TableHeadings→{{"Weights
→","Biases →"},None}]
Out[7]//TableForm=
```

$$\text{Weights} \rightarrow \begin{array}{l} -0.779 \\ 0.0435 \end{array}$$

$$\text{Biases} \rightarrow \begin{array}{l} 0. \\ 0. \end{array}$$

For instance, a layer can receive a length of 1 vector to produce an output vector of size 2.

```
In[8]:= LinearL[4]
Out[8]= {-3.11505,0.174007}
```

When input is not introduced in the appropriate shape, the layer will not be evaluated.

```
In[9]:= LinearL[{88,99}]
Out[9]= LinearLayer::invindata1: Data supplied to port "Input" was a
length-2 vector of real numbers, but expected a length-1 vector of real
numbers.
$Failed
```

The weights and biases are the parameters that the model must learn from, which can be adapted based on the input data that the model receives, which is why it is initialized randomly, since if we try to extract these values without initializing, we will not be able to since they have not been defined.

Layers have the property of being differentiable. This is achieved with NetPortGradient, which can represent the gradient of the output of a net respect to a port or to a parameter. For example, give the derivative of the output with respect to the input for a certain input value.

```
In[10]:= LinearL[2,NetPortGradient["Input"]]
Out[10]= {-0.735261}
```

Mean Squared Layer

Until now, we have seen the linear layer, which has various properties. Layers with the icon of a connected rhombus (Figure 9-8), by contrast, do not contain any learnable parameters, like MeanSquaredLossLayer, AppendLayer, SummationLayer, DotLayer, ContrastiveLossLayer, and SoftmaxLayer,among others.

```
In[11]:= MeanSquaredLossLayer[]
Out[11]:=
```

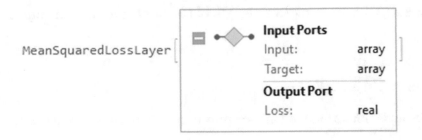

Figure 9-8. *MeanSquaredLossLayer*

MeanSquaredLossLayer[], has more than one input; that is because this layer computes the mean squared loss, which is the following expression

$\frac{1}{n}\sum(\text{Input} - \text{Target})^2$ and has the property that compares two numeric arrays. With the MeanSquaredLossLayer, the input/output ports' dimensions are entered in the same form as a linear layer, and the values of the input and target are entered as Associations.

```
In[12]:= MeanSquaredLossLayer["Input"→{3,2},"Target"→{3,2}][Association["
Input"→{{1,2},{2,1},{3,2}},"Target"→{{2,2},{1,1},{1,3}}]]
Out[12]= 1.6667
```

The latter example computes a MeanSquaredLossLayer for input/output dimensions of three rows and two columns or by defining first the layer and then applying the layer to the data.

Note   Use the command MatrixForm[{{1, 2}, {2, 1}, {3, 2}}] to verify the matrix shape of the data.

```
In[13]:= LossLayer= MeanSquaredLossLayer["Input"→{3,2},"Target"→{3,2} ];
LossLayer@
<|"Input"→{{1,2},{2,1},{3,2}},"Target"→{{2,2},{1,1},{1,3}}|>
Out[13]= 1.16667
```

To get more details about a layer (Figure 9-9), use the Information command.

```
In[14]:= Information[LossLayer]
Out[14]=
```

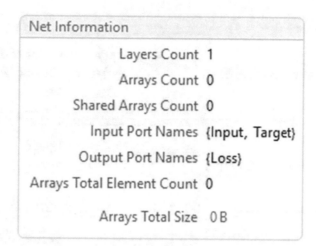

Figure 9-9. *Information about the loss layer*

To know the layer options, use the following.

```
In[15]:= MeanSquaredLossLayer["Input"→"Real","Target"→"Real"]//Options
Out[15]= {BatchSize→Automatic,NetEvaluationMode→Test,RandomSeeding→
Automatic,TargetDevice→CPU,WorkingPrecision→Real32}
```

The input port and target port options are similar to that of the linear layer with the different forms, Input→ "Real", n (form of a vector n), {n1 x n2 x n3} ... (an array of n dimensions), "Varying" (a vector or varying form) or a NetEncoder, but with the exception that the input and target must have the same dimensions. A few forms of layers are shown in Figure 9-10.

```
In[16]:= {MeanSquaredLossLayer["Input"→"Varying","Target"→"Varying"],Mean
SquaredLossLayer["Input"→ NetEncoder["Image"],"Target"→ NetEncoder["Image
"]],MeanSquaredLossLayer["Input"→1,"Target"→1]}//Dataset
Out[16]=
```

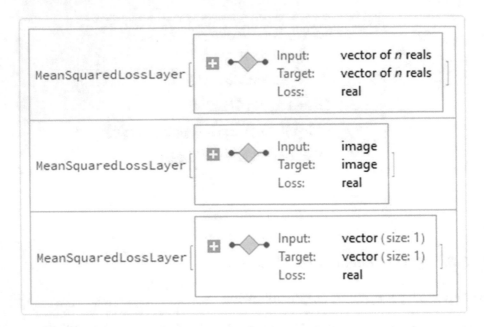

Figure 9-10. *Loss layers with different input and target forms*

Activation Functions

Activation functions are a crucial part for the construction of a neural network. The role of an activation function is to return an output from an established range, given an input. In the Wolfram Language activation function are treated as layers. The layer that is widely used in the Wolfram Language neural net framework is the ElementwiseLayer.

With this layer we can represent layers that can apply a unary function to the elements of the input data—in other words, a function that receives only one argument. These functions are also known as activation functions. For example, one of the most common functions used is the hyperbolic tangent (Tanh[x]), which is shown in Figure 9-11.

```
In[17]:= ElementwiseLayer[Tanh[#]&](* Altnernate form
ElementwiseLayer[Tanh]*)
Out[17]=
```

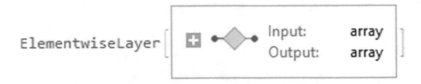

Figure 9-11. *Tanh[x] layer*

Elementwise layers do not have learnable parameters. The pure function is used because layers cannot receive symbols. If the plus icon is clicked, detailed information about the ports is shown as well as the parameters with the associated function, which in this case is Tanh. Having defined an ElementWiseLayer, it can receive values, like the other layers.

```
In[18]:= In[52]:= ElementwiseLayer[Tanh[#]&];
Table[%[i],{i,-5,5}]
Out[18]= {-0.999909,-0.999329,-0.995055,-0.964028,-
0.761594,0.,0.761594,0.964028,0.995055,0.999329,0.999909}
```

When no input or output shape is given, the layer will infer the type of data it will receive or return. For instance, by specifying only the input as real, Mathematica will infer that the output will be real (Figure 9-12).

```
In[19]:= TanhLayer=ElementwiseLayer[Tanh,"Input"→ "Real"]
Out[19]=
```

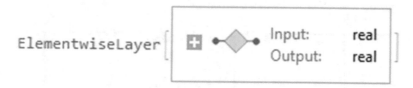

Figure 9-12. *ElementWiseLayer with the same output as the input*

Or, this can be inferred by entering only the output (Figure 9-13) for a Rectified Linear Unit (ReLU).

```
In[20]:= RampLayer=ElementwiseLayer[Ramp,"Output"→ {1}](*or ElementwiseLay
er["ReLU","Output"\[Rule] "Varying"]*)
Out[20]=
```

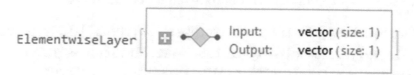

Figure 9-13. *Ramp function or ReLU*

> **Note**   Clicking on the plus icon will show the established function of the ElementWiseLayer, as well as the details of the layer ports.

Every layer in the Wolfram Language can be run through a graphics processor unit (GPU) or a central processing unit (CPU), by specifying the TargetDevice option. For example let's plot the previously created layers with the TargetDevice on the CPU (Figure 9-14).

```
In[21]:= GraphicsRow@{Plot[TanhLayer[x,TargetDevice→
"CPU"],{x,-12,12},PlotLabel→"Hiperbolic
Tangent",AxesLabel→{Style["x",Bold],Style["f(x)",Italic]},PlotStyle→
ColorData[97,25],Frame→True],Plot[RampLayer[x,TargetDevice→ "CPU"],{x,-
12,12},PlotLabel→"ReLU",AxesLabel→{Style["x",Bold],Style["f(x)",Italic]},
PlotStyle→ColorData[97,25],Frame→True]}
Out[21]=
```

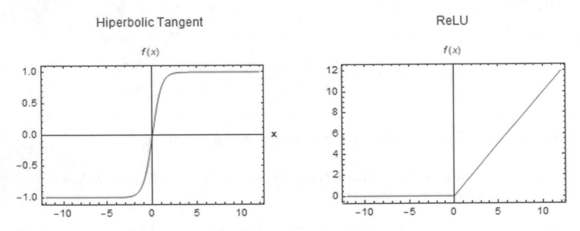

Figure 9-14. *Tanh[x] and Ramp[x] activation functions*

Other functions can be used by their name or by Wolfram Language syntax—for instance, the SoftPlus function. This is demonstrated in Figure 9-15.

```
In[22]:= GraphicsRow@{Plot[ElementwiseLayer["SoftPlus"][x,TargetDevice→
"CPU"],{x,-12,12},PlotLabel→#1,AxesLabel→#2,PlotStyle→#3,Frame→#4],
Plot[ElementwiseLayer[Log[Exp[#]+1]&][x,TargetDevice→ "CPU"],{x,-12,12},
PlotLabel→#1,AxesLabel→#2,PlotStyle→#3,Frame→#4]}&["SoftPlus",
{Style["x",Bold],Style["f(x)",Italic]},ColorData[97,25],True]
Out[22]=
```

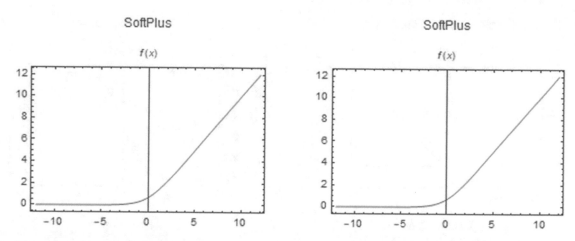

Figure 9-15. *SoftPlus function generated by the associated name and pure function*

Other common functions are shown in the next plots, such as the scaled exponential linear unit, sigmoid, hard sigmoid, and hard hyperbolic tangent (Figure 9-16). To view the functions supported, visit the documentation and type ElementwiseLayer in the search box.

```
In[23]:=
GraphicsGrid@{{Plot[ElementwiseLayer["ScaledExponentialLinearUnit"]
[i,TargetDevice→ #1],#2,AxesLabel→#3,PlotStyle→#4,Frame→#5,PlotLabel→
"ScaledExponentialLinearUnit"],
Plot[ElementwiseLayer[LogisticSigmoid][i,TargetDevice→ #1],#2,AxesLabel→
#3,PlotStyle→#4,Frame→#5,PlotLabel→"LogisticSigmoid"]},
{Plot[ElementwiseLayer["HardSigmoid"][i,TargetDevice→ #1],#2,AxesLabel→
#3,PlotStyle→#4,Frame→#5,PlotLabel→"HardSigmoid"],
Plot[ElementwiseLayer["HardTanh"][i,TargetDevice→ #1],#2,AxesLabel→#3,
PlotStyle→#4,Frame→#5,PlotLabel→"HardTanh"]}}&["CPU",{i,-10,10},
{Style["x",Bold],Style["f(x)",Italic]},ColorData[97,25],True]
Out[23]=
```

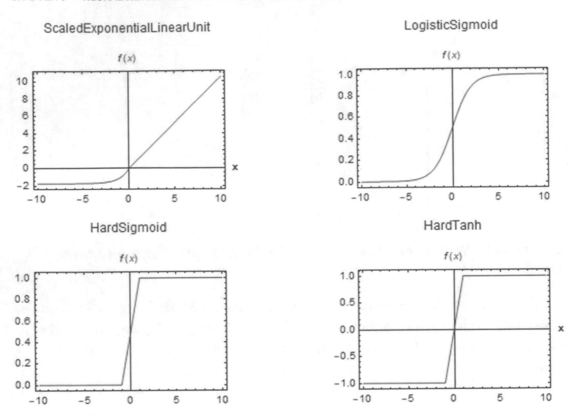

Figure 9-16. *Plot of four different activation functions*

SoftmaxLayer

SoftmaxLayer is a layer that uses the expression $S(x_i) = \dfrac{\exp(x_i)}{\sum_{j=1}^{n} \exp(x_j)}$, where x represents a vector and x_i the components of the vector. This expression is known as the Softmax function. The functionality of this layer consists of converting a vector to a normalized vector, which consists of values in the range of 0 to 1. This layer is generally used to represent a partition of the classes based on the probabilities of each one, and it is used for tasks that involve classification.

The input and output forms in the SoftmaxLayer can be entered as the other common layers except for the shape of "Real."

```
In[24]:= SFL=SoftmaxLayer["Input"→ 4,"Output"→ 4];
```

Now the layer can be applied to data.

```
In[25]:= SetAccuracy[SFL[{9,8,7,6}],3]
Out[25]= {0.64,0.24,0.09,0.03}
```

The total of the latter equals 1. SoftmaxLayer allows us to specify the level depth of normalization. This is seen in the parameter's properties of the layer. A level of -1 produces the normalization of a flatten list. Also, SoftmaxLayer can receive multidimensional arrays, not just flatten lists.

```
In[26]:= SoftmaxLayer[1,"Input"→{3,2}];
SetPrecision[%[{{7,8},{8,7},{7,8}}],3]//MatrixForm
Out[26]//MatrixForm=
```

$$\begin{pmatrix} 0.212 & 0.422 \\ 0.576 & 0.155 \\ 0.212 & 0.422 \end{pmatrix}$$

Summing the elements of the first columns gives one the same for the second column.

Another practical layer is called CrossEntropyLossLayer. This layer is widely used as a loss function for classification tasks. This loss layer measures how well the classification model performs. Entering the string Probabilities as argument of the loss layer computes the cross-entropy loss by comparing the input class probability to the target class probability.

```
In[27]:= CrossEntropyLossLayer["Probabilities","Input"→3];
```

Now the target form is set to the probabilities of the classes; the inputs and targets are entered in the same way as with MeansSquaredLoss.

```
In[28]:= %[<|"Input"→{0.2,0.5,0.3},"Target"→{0.3,0.5,0.2}|>]
Out[28]= 1.0702
```

Setting the Binary argument in the layer is used when the probabilities constitutes a binary alternative.

```
In[29]:=CrossEntropyLossLayer["Binary","Input"→ 1];
%[<|"Input"→ 0.1,"Target"→ 0.9|>]
Out[29]= 2.08286
```

To summarize the properties of layers in the Wolfram Language, the inputs and outputs of the layers are always scalars and numeric matrices. Layers are evaluated using lower number precision, such as single precision numbers. Layers have the property of being differentiable, this helps the model to perform efficient learning, since some learning methods go into convex optimization problems.

Within the Wolfram Language, there are many layers, each with specific functions. To display all the layers that are within Mathematica, it is advisable to check the documentation or write ?* Layer, which will give us the commands that have the word layer associated at the end. Each layer has different behaviors, operations, and parameters, although some may resemble other commands such as Append and AppendLayer. It is important to know the different layers and what they are capable of doing to make the best use of it.

Encoders and Decoders

Encoders

If audio, images, or other types of variables are intended to be used, this type of data needs to be converted into a numeric array in order to be introduced as input into a layer. Layers must have a NetEncoder attached to the input in order to perform a correct construction. The NetEncoders interpret the image, audio, and so forth, data to a numeric value in order to be used inside a net model. Different names are associated with the encoding type. The most common are: Boolean (True or False, encoding as 1 or 0), Characters (string characters as on-hot vector encoding), Class (class labels as integer encoding), Function (custom function encoding), Image (2D image encoding as a rank 3 array), and Image3D (3D image encoding as a rank 4 array).

The arguments of the encoder are the name or the name and the corresponding features of the encoder (Figure 9-17).

```
In[30]:= NetEncoder["Boolean"]
Out[30]=
```

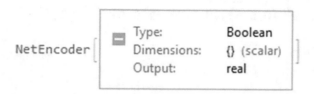

Figure 9-17. *Boolean type NetEncoder*

To test the encoder, we use the following.

```
In[31]:= Print["Booleans:",{%[True],%[False]}]
Booleans:{1,0}
```

A NetEncoder can have classes with different index labels. Like a classification of a class X and class Y, this will correspond to an index of the range from 1 to 2 (Figure 9-18).

```
In[32]:= NetEncoder[{"Class",{"Class X","Class Y"}}]
Out[32]=
```

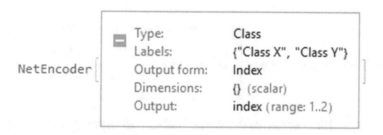

Figure 9-18. *Class type NetEncoder*

```
In[33]:= Print["Classes:",%[Table[RandomChoice[{"Class X","Class
Y"}],{i,10}]]]
Classes:{2,1,2,2,1,1,2,1,1,1}
```

The following is used for a unit vector.

```
In[34]:= NetEncoder[{"Class",{"Class X","Class Y","Class Z"},"UnitVector"}];
Print["Unit Vector:",%[Table[RandomChoice[{"Class X","Class Y","Class
Z"}],{i,5}]]]
Print["MatrixForm:",%%[Table[RandomChoice[{"Class X","Class Y","Class
Z"}],{i,5}]]//MatrixForm[#]&]
```

Unit Vector:{{0,1,0},{0,0,1},{0,1,0},{1,0,0},{1,0,0}}

$$
\text{MatrixForm:} \begin{pmatrix} 0 & 0 & 1 \\ 0 & 0 & 1 \\ 1 & 0 & 0 \\ 0 & 1 & 0 \\ 1 & 0 & 0 \end{pmatrix}
$$

Depending on the name used inside NetEncoder, properties related to the encoder may vary. This is depicted in different encoder objects created. To attach a NetEncoder to a layer, the encoders are entered at the input port—for example, for a ElementwiseLayer (Figure 9-19). In this case, the input port of the layer has the name Boolean; the layer recognizes that this is a NetEncoder of a Boolean type. Clicking on the name Boolean will show the relevant properties.

```
In[35]:= ElementwiseLayer[Sin,"Input"→NetEncoder["Boolean"]]
Out[35]//Short=
```

Figure 9-19. *Layer with an encoder attached to the input port*

For a LinearLayer, use the following form.

```
In[36]:= LinearLayer["Input"→NetEncoder[{"Class",{"Class X","Class Y"}}],
"Output"→ "Scalar"]
Out[36]=
```

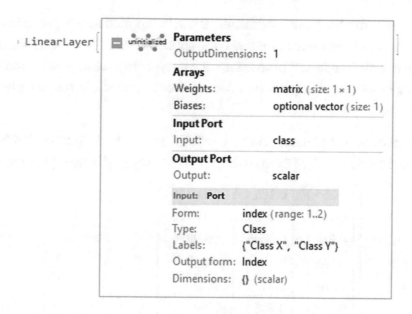

Figure 9-20. *Class encoder attached to a Linear Layer*

Clicking on the input port will show the encoder specifications, as Figure 9-20 shows.

A NetEncoder is also used to convert images into numeric matrices or arrays, by specifying the class, the size or width, and height of the output dimensions, and the color space, which can be grayscale, RGB, CMYK, or HSB (hue, saturation, and brightness)—for example, encoding an image that produces a 1 x 28 x 28 array in grayscale, or 3 x 28 x 28 array in an RGB scale (Figure 9-21), no matter the size of the input image. The first rank of the array represents the color channel, and the other two represent the spatial dimensions.

```
In[37]:= Table[NetEncoder[{"Image",{28,28},"ColorSpace"→ Color}],{Color,
{"Grayscale","RGB"}}]
Out[37]=
```

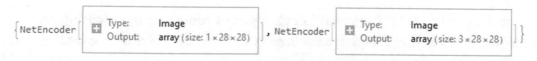

Figure 9-21. *NetEncoders for grayscale and RGB scale images*

Once the encoder has been established, it can be applied to the desired image, then the encoder creates a numeric matrix with the specified size. Creating a NetEncoder for an image will show relevant properties such as type, input image size, and color space, among others. Applying the encoder will create a matrix in the size previously established.

```
In[38]:= ImgEncoder=NetEncoder[{"Image",{3,3},"ColorSpace"→ "CMYK"}];
Print["Numeric Matrix:",%[ExampleData[{"TestImage","House"}]]//MatrixForm]
```

Numeric Matrix:

$$
\begin{pmatrix}
\begin{pmatrix} 0.255 \\ 0.168 \\ 0.255 \end{pmatrix} & \begin{pmatrix} 0.145 \\ 0.00392 \\ 0.0235 \end{pmatrix} & \begin{pmatrix} 0.0784 \\ 0.0118 \\ 0.0274 \end{pmatrix} \\
\begin{pmatrix} 0.153 \\ 0.196 \\ 0.129 \end{pmatrix} & \begin{pmatrix} 0.2 \\ 0.31 \\ 0.255 \end{pmatrix} & \begin{pmatrix} 0.259 \\ 0.349 \\ 0.306 \end{pmatrix} \\
\begin{pmatrix} 0.047 \\ 0.102 \\ 0.00784 \end{pmatrix} & \begin{pmatrix} 0.164 \\ 0.321 \\ 0.164 \end{pmatrix} & \begin{pmatrix} 0.262 \\ 0.384 \\ 0.176 \end{pmatrix} \\
\begin{pmatrix} 0.16 \\ 0.262 \\ 0.184 \end{pmatrix} & \begin{pmatrix} 0.255 \\ 0.408 \\ 0.478 \end{pmatrix} & \begin{pmatrix} 0.325 \\ 0.388 \\ 0.569 \end{pmatrix}
\end{pmatrix}
$$

The output generated is a numeric matrix that is now ready to be implemented in a network model. If the input image shape is in a different color space, the encoder would reshape and transform the image into the established color space The image use in this example is obtained from the ExampleData[{"TestImage","House"}].

Pooling Layer

Encoders can be added to the ports of single layers or containers by specifying the encoder to the port—for instance, a PoolingLayer. These layers are used mostly in convolutional neural networks (Figure 9-22).

```
In[39]:= PoolLayer=PoolingLayer[{3,3},{2,2},PaddingSize→0,"Function"→
Max,"Input"→ NetEncoder[{"Image",{3,3},"ColorSpace"→ "CMYK"}](*Or
ImgEncoder*)]
Out[39]=
```

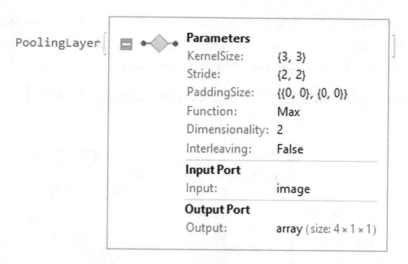

Figure 9-22. *PoolingLayer with a NetEncoder*

The latter layer has a specification for a two-dimensional PoolingLayer with a kernel size of 3 x 3 and a stride of 2 x 2, which is the step size between kernel applications. PaddingSize adds elements at the beginning and the end of the matrix. This is used so that the division between the matrix and the kernel size is an integer number, in order to avoid losing information between layers. Function indicates the pooling operation function, which is Max; with this, it calculates the maximum value in each patch of each filter, the mean for the average value of the filter, and the total for the summation of the values of the filter. Sometimes they might be known as max and average pooling layers.

```
In[40]:= PoolLayer[ExampleData[{"TestImage","House"}]]//MatrixForm
Out[40]//MatrixForm=
```

$$\begin{pmatrix} (0.255) \\ (0.349) \\ (0.384) \\ (0.569) \end{pmatrix}$$

Decoders

Once the operations of the net are finished, it will return numeric expressions. On the other hand, in some tasks, we do not want numeric expressions, such as in classification tasks where classes can be given as outputs, where the model is able to tell that a certain object belongs to a class A and another object belongs to a class B, so a vector or numeric array can represent a probability of each class. In order to convert the numeric arrays into other forms of data, a NetDecoder is used (Figure 9-23).

```
In[41]:= Decoder=NetDecoder[{"Class",CharacterRange["W","Z"]}]
Out[41]=
```

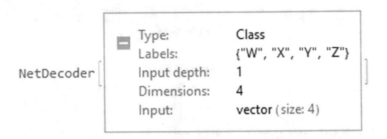

	Type:	Class
NetDecoder	Labels:	{"W", "X", "Y", "Z"}
	Input depth:	1
	Dimensions:	4
	Input:	vector (size: 4)

Figure 9-23. *NetDecoder for four different classes*

The dimension of the decoder is equal to class construction. We can apply a vector of probabilities, and the decoder will interpret it and tell us the class to which it belongs. It will also display the probabilities of the classes.

```
In[42]:= Decoder@{0.3,0.2,0.1,0.4}(*This is the same as Decoder[{0.3,0.2,0.
1,0,4},"Decision"] *)
Out[42]= Z
```

TopDecisions, TopProbabilites, and uncertainty of the probability distribution are displayed as follows.

```
In[43]:= TableForm[{Decoder[{0.3,0.2,0.1,0.4},"TopDecisions"→ 4](* or
{"TopDecisions", 4} the same is for TopProbabilities*),
Decoder[{0.3,0.2,0.1,0.4},"TopProbabilities"→ 4],
Decoder[{0.3,0.2,0.1,0.4},"Entropy"]},TableDirections→Column,TableHeadings
→{{Style["TopDecisions",Italic],Style["TopProbabilities",Italic],
Style["Entropy",Italic]},None}]
Out[45]//TableForm=
```

TopDecisions	Z	W	X	Y
TopProbabilities	Z→0.4	W→0.3	X→0.2	Y→0.1
Entropy	1.27985			

Input depth is added to define the level of application of the class given the list of values.

```
In[44]:= NetDecoder[{"Class",CharacterRange["X","Z"],"InputDepth"→2}];
```

Applying the decoder to a nested list of values will produce the following.

```
In[45]:= TableForm[{%[{{0.1,0.3,0.6},{0.3,0.4,0.3}},"TopDecisions"→ 3]
(* or {"TopDecisions", 4} the same is for TopProbabilities*),
%[{{0.1,0.3,0.6},{0.3,0.4,0.3}},"TopProbabilities"→ 3],
%[{{0.1,0.3,0.6},{0.3,0.4,0.3}},"Entropy"]},TableDirections→Column,Table
Headings→{{Style["TopDecisions",Italic],Style["TopProbabilities",Italic],
Style["Entropy",Italic]},None}]
Out[45]//TableForm=
```

	Z	Y
TopDecisions	Y	X
	X	Z
	Z→0.6	Y→0.4
TopProbabilities	Y→0.3	X→0.3
	X→0.1	Z→0.3
Entropy	0.897946	1.0889

A decoder is added to the output port of a layer, container, or a network model.

```
In[46]:=SoftmaxLayer["Output"→NetDecoder[{"Class",{"X","Y","Z"}}]];
```

Applying the layer to the data will produce the probabilities for each class.

```
In[47]:= {%@{1,3,5},%[{1,3,5},"Probabilities"],%[{1,3,5},"Decision"]}
Out[47]={Z,<|X→0.0158762,Y→0.11731,Z→0.866813|>,Z}
```

Applying Encoders and Decoders

We are ready to implement the whole process of encoding and decoding in Figure 9-24. First, the image will be resized by 200 pixels in width to show how the original image looks before encoding.

```
In[48]:= Img=ImageResize[ExampleData[{"TestImage","House"}],200]
Out[48]=
```

Figure 9-24. *Example image of a house*

Then the encoder and decoder are defined.

```
In[49]:=
Encoder=NetEncoder[{"Image",{100,100},"ColorSpace"→ "RGB"}];
Decoder=NetDecoder[{"Image",ColorSpace→ "Grayscale"}];
```

Then the encoder is applied to the image, and the decoder is applied to the numeric matrix. The dimensions of the decoded image are checked to see if they match the encoder output dimensions.

```
In[50]:=
Encoder[Img];
Decoder[%]
```

Figure 9-25. *Example image of a house*

As seen in Figure 9-25, the image has been converted into a grayscale image with new dimensions.

```
In[51]:= ImageDimensions[%]
Out[51]= {100,100}
```

As seen, the picture has been resized. Try and have a look the steps in the process, like viewing the numeric matrix and the objects corresponding to the encoder and decoder. The use of the encoders and decoders involves the type of data you are using because every net model receives different inputs and generates different outputs.

NetChains and Graphs
Containers

Neural networks consist of different layers, not individual layers on their own. To construct more complex structures that are more than one layer, the command NetChain or NetGraph is used. These containers are valuable to properly operate and construct neural networks in the Wolfram Language. In the Wolfram Language, containers are structures that assemble the infrastructure of the neural network model. Containers can have multiple forms. NetChain is useful to create linear and non-linear structures' nets. This helps the model to learn non-linear patterns. We can think that each layer that exists in a network will have a level of abstraction that serves to detect complex behavior, which could not be recognized if we only worked with one single layer. As a result, we can build networks in a general way, starting from three layers: input layer, hidden layer, and output layer. When we have more than two hidden layers, we are talking about the term Deep Learning; for more reference visit *Introduction to Deep Learning: from logical calculus to artificial intelligence* by Sandro Skansi (2018: Springer).

NetChain can join two operations. They can be written as a pure function, instead of just the name of the function (Figure 9-26).

```
In[52]:= NetChain[{ElementwiseLayer[LogisticSigmoid@#&],ElementwiseLayer
[Sin@#&]}]
Out[52]=
```

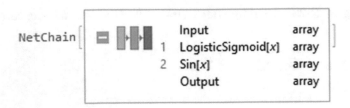

Figure 9-26. *NetChain containing two elementwise layers*

The object returned is a NetChain, and the icon of three colored rectangles appears. This means that the object created (NetChain) or referred is a net chain and contains layers. If the chain is examined, it will show the input, first (LogisticSigmoid), second (Sin), and output layers. The operations are in order of appearance, so the first layer is applied and then the second. The input and output options of other layers are supported in NetChain, such as a single real number (Real), an integer (Integer), an "n"-length vector, and a multidimensional array.

```
In[53]:= NetInitialize@NetChain[{3,4,12,Tanh},"Input"→1]
Out[53]=
```

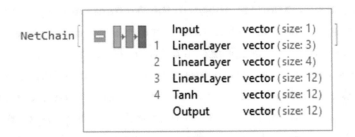

Figure 9-27. *NetChain with multiple layers*

NetChain recognizes the Wolfram Language function names and associates them with their corresponding layers, like 3, 4, and 12. They represent a linear layer with outputs of the size 3, 4, and 12 (Figure 9-27). The function Tanh represents the elementwise layer.

Let's append a layer to the chain created with NetAppend (Figure 9-28) or NetPrepend. If you notice, many of the original commands of the Wolfram Language have the same meaning—for example, to delete in a chain would be NetDelete[net_ name, #_of_layer].

```
In[54]:= NetInitialize@NetChain[{1,ElementwiseLayer[LogisticSigmoid@#&]},
"Input"→ 1];
NetCH2=NetInitialize@NetAppend[%,{1,ElementwiseLayer[Cos[#]&]}]
Out[54]=
```

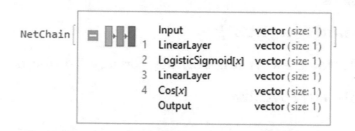

Figure 9-28. *NetChain object with the added layers*

When a net is applied to data, different options are available, such as NetEvaluationMode (mode of evaluation either train or test), TargetDevice, and WorkingPrecision (numeric precision).

```
In[55]:= NetCH2[{{0},{2},{44}},NetEvaluationMode→ "Train",TargetDevice→
"CPU",WorkingPrecision→ "Real64",RandomSeeding→ 8888](*use N@Cos[Sin
[LogisticSigmoid[{0,2,44}]]] to check results*)
Out[55]= {{0.919044},{0.991062},{1.}}
```

Another form is to enter the explicit names of layers in a chain. This are typed as an association (Figure 9-29).

```
In[56]:=NetInitialize@NetChain[<|"Linear Layer 1"→LinearLayer[3], "Ramp"→
Ramp,"Linear Layer 2"→LinearLayer[4],"Logistic"→ ElementwiseLayer[Logisti
cSigmoid]|>,"Input"→ 3]
Out[56]=
```

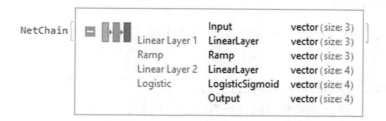

Figure 9-29. *NetChain object with costumed layer names*

Inspecting the contents of the layer should appear after clicking the name of the layer or the layer.

If a layer wants to be extracted, then NetExtract is used along with the name of the corresponding layer. The output is suppressed, but the layer should pop out if the semicolon is removed.

```
In[57]:=NetExtract[%,"Logistic"];
```

To extract all of the layers in one line of code, Normal will do the job (Figure 9-30).

```
In[58]:= Normal[NetCH2]//Column
Out[58]=
```

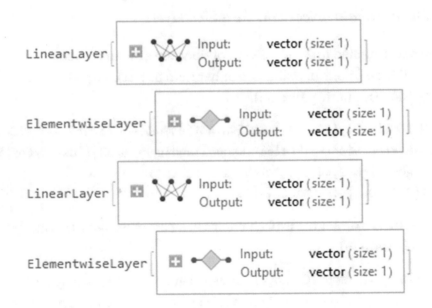

Figure 9-30. *Layers of the NetChain NetCH2*

Multiple Chains

Chains can be joined together, as with a nested chain (Figure 9-31).

```
In[59]:=
chain1=NetChain[{12,SoftmaxLayer[]}];
chain2=NetChain[{1,ElementwiseLayer[Cos[#]&]}];
NestedChain=NetInitialize@NetChain[{chain1,chain2},"Input"→ 12]
Out[59]=
```

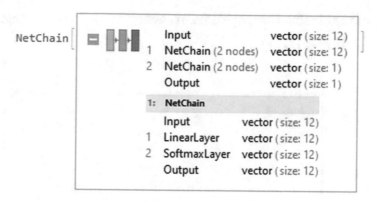

Figure 9-31. *Chain 1 selected of the two chains available*

This chain is divided in two NetChains, where each chain represents a chain. In this case, we see chain1 and chain2, and each chain shows its corresponding nodes. To flatten the chains, use NetFlatten (Figure 9-32).

```
In[60]:= NetFlatten[NestedChain]
Out[60]=
```

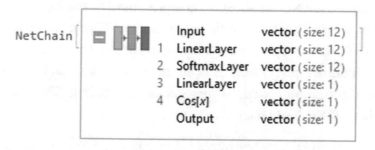

Figure 9-32. *Flattened chain*

NetGraphs

The command NetChain only joins layers in which the output of a layer is connected to the input of the next layer. NetChain does not work in connecting inputs or outputs to other layers; it only works with one layer. To work around this, the use of NetGraph is required. Besides allowing more inputs and layers, NetGraph represents the structure and the process of the neural network with a graph (Figure 9-33).

```
In[62]:= NetInitialize@NetGraph[{ LinearLayer["Output"→ 1,"Input"→
1],Cos,SummationLayer[]},{}]
Out[61]=
```

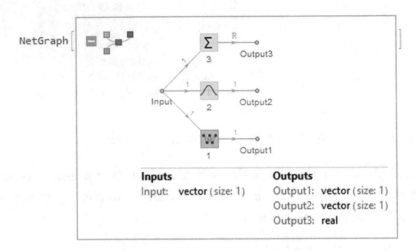

Figure 9-33. *Expanded NetGraph*

The object crafted is a NetGraph, and it is represented by the figure of the connecting squares. As seen in Figure 9-33. The input goes to three different layers, and each layer has its output. NetGraph accepts two arguments: the first is for the layers or chains, and the second is to define the graph vertices or connectivity of the net. For example, in latter code the net has three outputs because the vertices were not specified. SummationLayer is a layer that sums all the input data.

```
In[62]:= Net1=NetInitialize@NetGraph[{ LinearLayer["Output"→ 2,"Input"→
1],Cos,SummationLayer[]},{1→ 2→ 3}]
Out[62]=
```

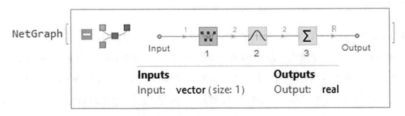

Figure 9-34. *Unidirectional NetGraph*

The vertex notation means that the output of a layer is given to another layer, and so on. In other words, $1 \rightarrow 2 \rightarrow 3$ means that the output of the Linear Layer is passed to the next layer until it is finally summed in the last layer with SummationLayer (Figure 9-34). Thus preserving the order of appereance of the layers, however we can alter the order of each vertex.

The net can be modified so that outputs can go to other layers of the net, such as 1 to 3 and then to 2 (Figure 9-35). With NetGraph, layers and chains can be entered as a list or an association. The vertices are typed as a list of rules.

```
In[63]:= Net2=NetInitialize@NetGraph[{ LinearLayer["Output"→ 2,"Input"→ 1],
Cos,SummationLayer[]},{1→ 3→2}]
Out[63]=
```

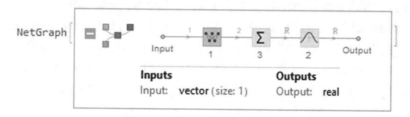

Figure 9-35. *NetGraph structure*

The inputs and outputs of each layer are marked by a tooltip that appears when passing the cursor over the graph lines or vertices. In the sense that input and output are not specified, NetGraph will infer the type of data in the input and output port; this is the case for the capital R in the input and output of the ElementwiseLayer, which stands for real.

With NetGraph, layers can be entered as a list or as an association. The connections are typed as a list of rules (Figure 9-36).

```
In[64]:= NetInitialize@NetGraph[<|"Layer 1"→ LinearLayer[2,"Input"→
1],"Layer 2"→ Cos,"Layer 3"→ SummationLayer[]|>,{"Layer 2"→ "Layer 1"→
"Layer 3"}]
Out[64]=
```

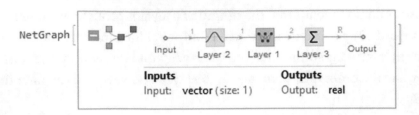

Figure 9-36. *NetGraph with named layers*

Now, it is possible to specify how many inputs and outputs a structure can have from the NetPort command (Figure 9-37).

```
In[65]:= NetInitialize@NetGraph[{ LinearLayer[3,"Input"→ 1],
LinearLayer[3,"Input"→ 2], LinearLayer[3,"Input"→ 1],
TotalLayer[]},{NetPort["1st Input"]→ 1, NetPort["2nd Input"]→ 2,
NetPort["3rd Input"]→ 3,{1,2,3}→ 4}](*Or NetInitialize@NetGraph[<|"L1"→
LinearLayer[3,"Input"→,"L2"→ LinearLayer[3,"Input"→1], "L3"\[Rule]
LinearLayer[3,"Input"→ 1],"Tot L"→TotalLayer[]|>,{NetPort["1st Input"] →
"L1", NetPort["2nd Input"] → "L2",NetPort["3rd Input"] → "L3",
{"L1","L2","L3"} → "Tot L"}]*)
Ouy[65]=
```

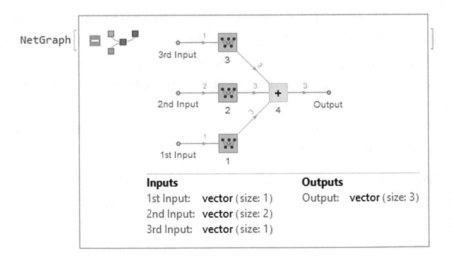

Figure 9-37. *NetGraph with multiple inputs*

In the event that we have more than one input, each input is entered in the specified port.

```
In[66]:= %[<|"1st Input"-> 32.32,"2nd Input"-> {2,\[Pi]},"3rd Input"-> 1|>]
Out[66]= {-41.7285,-11.2929,19.0044}
```

Having more than one output, the results are displayed for every different output (Figure 9-38).

```
In[67]:= NetInitialize[NetGraph[{LinearLayer[1,"Input"→
1],LinearLayer[1,"Input"→ 1],LinearLayer[1,"Input"→ 1],Ramp,El
ementwiseLayer["ExponentialLinearUnit"],LogisticSigmoid},{1→4→
NetPort["Output1"],2→5→ NetPort["Output2"],3→ 6→ NetPort["Output3"]}],R
andomSeeding→8888]
Out[67]=
```

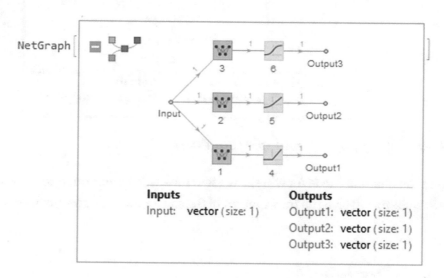

Figure 9-38. *NetGraph with three outputs*

```
Out[68]= <|Output1→{0.},Output2→{2.17633},Output3→{0.372197}|>
```

NetChain containers can be treated as layers with NetGraph (Figure 9-39). Some layers, such as the CatenateLayer, accept zero arguments.

```
In[69]= NetInitialize@NetGraph[{
LinearLayer[1,"Input"→ 1],
NetChain[{LinearLayer[1,"Input"→ 1],ElementwiseLayer[LogisticSigmoid[#]&]}],
NetChain[{LinearLayer[1,"Input"→ 1],Ramp}],
ElementwiseLayer["ExponentialLinearUnit"],
CatenateLayer[]
},{1→4,2→5,3→ 5,4→ 5}]
Ou[69]=
```

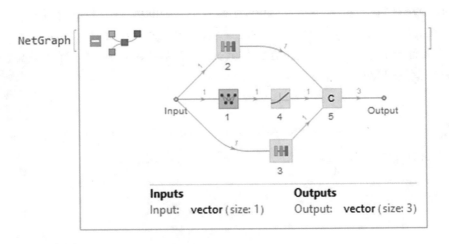

Figure 9-39. *NetGraph with multiple containers*

Clicking the chain or the layer will show the relevant information, and clicking the layer inside a chain will give the information of the layer contained on the selected chain.

Combining Containers

With NetGraph, NetChains and NetGraphs can be nested to form different structures, as seen in the next example (Figure 9-40), where a NetChain can be followed by a NetGraph and vice versa.

```
In[70]N1=NetGraph[{1,Ramp,2,LogisticSigmoid},{1→ 2,2→ 3,3→ 4}];
N2=NetChain[{3,SummationLayer[]}];
NetInitialize@NetGraph[{N2,N1},{2→ 1},"Input"→ 22]
Ouy[70]=
```

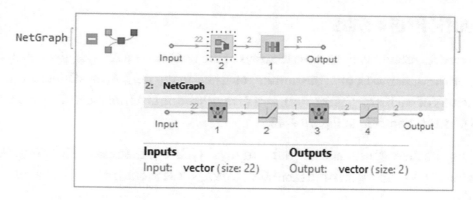

Figure 9-40. *Nested NetGraph*

From the graph in Figure 9-40, it is clear the input goes to the NetGraph and the output if the NetGraph goes to the NetChain. A NetChain or NetGraph that has not been initialized will appear in red. A fundamental quality of the containers (NetChain, NetGraph) is that they can behave as a layer. With this in mind, we can create nested containers involving only NetChains, NetGraphs, or both.

Just as a demonstration, more complex structures can be created with NetGraph like the in Figure 9-41. Once a network structure is created, properties about every layer or chain can be extracted. For instance, with "SummaryGraphic," you can obtain the graphic of the network graph.

```
In[71]:=Net=NetInitialize@NetGraph[{LinearLayer[10],Ramp,10,SoftmaxLayer[],
TotalLayer[],ThreadingLayer[Times]},{1→2→3→4,{1,2,3}→5,{1,5}→6},"Input
"→"Real"];
Information[Net,"SummaryGraphic"]
Out[71]=
```

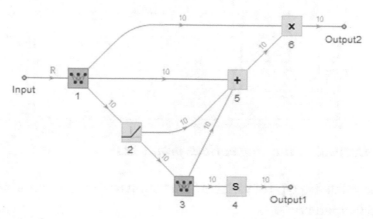

Figure 9-41. *Compound net structure*

Network Properties

The properties related to the numeric arrays of the network are Arrays (gives each array in the network), ArraysCount (the number of arrays in the net), ArraysDimensions (dimensions of each array in the net), and ArraysPositionList (position of each array in the net). This is depicted in Figure 9-42.

```
In[72]:= {Dataset@Information[Net,"Arrays"],Dataset@Information[Net,"Arrays
Dimensions"],Dataset@Information[Net,"ArraysPositionList"]}//Dataset
Out[72]=
```

Figure 9-42. *Datasat containing various properties*

Information related to type of variable in the input and output ports are shown with InputPorts and OutputPorts.

```
In[73]:= {Information[Net,"InputPorts"],Information[Net,"OutputPorts"]}
Out[73]= {<|Input→Real|>,<|Output1→10,Output2→10|>}
```

We can see that the input is a real number, and the net has two outputs vectors of size 10. The most used properties related to layers are Layers (returns every layer of the net), LayerTypeCounts (number of occurrence of a layer in the net), LayersCount (number of layers in the net), LayersList (a list of all the layers in the net), and LayerTypeCounts (number of occurrence of a layer in the net). Figure 9-43 shows for Layers and LayerTypeCounts.

```
In[74]:= Dataset@{Information[Net,"Layers"],Information[Net,"LayerTypeCounts"]}
Out[74]=
```

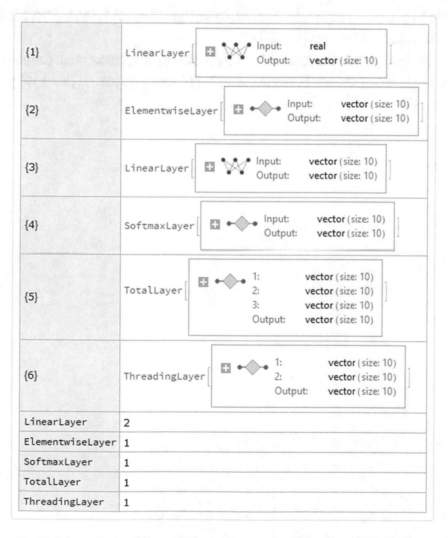

Figure 9-43. *Information about the layers contained in the symbol Net*

Visualization of the net structure (Figure 9-44) is achieved with the properties LayerGraph (a graph showing the connectivity of the layers), SummaryGraphics (graphic of the net structure), MXNetNodeGraph (MXNeT raw graph operations), and MXNetNodeGraphPlot (annotated graph of MXNet operations). MXNet is an open-source Deep Learning framework that supports a variety of programming languages, and one of them is the Wolfram Language. In addition, the Wolfram Neural Network Framework works with MXNet structure as backend support.

```
In[75]:= Grid[{{Style["Layers Connection",Italic,20,ColorData[105,4]],
Style["NetGraph",Italic,20,ColorData[105,4]]},{Information[Net,"LayersGraph
"],Information[Net,"SummaryGraphic"]},
{Style["MXNet Layer Graph",Italic,20,ColorData[105,4]],Style["MXNet Ops
Graph",Italic,20,ColorData[105,4]]},
{Information[Net,"MXNetNodeGraph"],Information[Net,"MXNetNodeGraphPlot"]}},
Dividers→All,Background→{{{None,None}},{{Opacity[1,Gray],None}}}]
Normal@Keys@%
Out[75]=
```

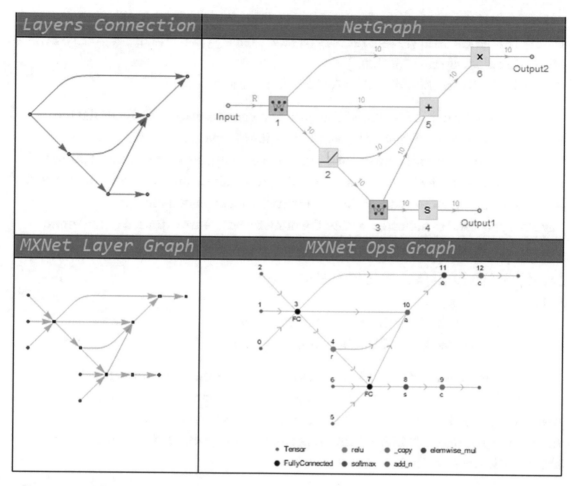

Figure 9-44. *Grid showing multiple graphics*

Passing the cursor pointer over a layer or node in the MXNet symbol graph, a tooltip is displayed showing the properties of the MXNet symbols like ID, name, parameters, attributes, and inputs.

Exporting and Importing a Model

Because of the interoperability of the Wolfram Language and MXNet, the Wolfram Language supports the import and export of neural nets, initialized or uninitialized. For this, we create a folder on the desktop with the MXNet Nets name. We will export the network found in the Net variable.

```
In[76]:= FileDirectory="C:\\Users\\My-pc\\Desktop\\MXNet Nets\\";
Export[FileNameJoin[{FileDirectory,"MxNet.json"}],Net,"MXNet","ArrayPath"→
Automatic,"SaveArrays"→ True]
Out[76]= C:\Users\My-pc\Desktop\MXNet Nets\MxNet.json
```

Exporting the network to the MXNet format generates two files: a JSON file that
stores the topology of the neural network and a file of type .params that contains
the required parameters (numeric arrays used in the network) data for the exported
architecture, once it has been initialized. With ArrayPath set on Automatic, the params
file is saved in the same folder of the net, otherwise it can have a different path.
SaveArrays is used to indicate whether the numeric arrays are exported (True) or not
(False). Let us check the two files created in the MXNets Nets folder.

```
In[77]:= FileNames[All,File@FileDirectory]
Out[77]=
{C:\Users\My-pc\Desktop\MXNet Nets\MxNet.json,
C:\Users\My-pc\Desktop\MXNet Nets\MxNet.params}
```

To import an MXNet network, the JSON file and .params is recommended to be in
the same folder, because the Wolfram Language will assume that a certain JSON file will
match the pattern of the .params file. There are various ways to import a net, including
Import[file_name.json,"MXNet"] and Import[file_name.json,{"MXNet",element}] (the
same as with .param files).

```
In[78]:= Import[FileNameJoin[{FileDirectory,"MxNet.json"}],{"MXNet","Net"},
"ArrayPath"→ Automatic];
```

The latter net was imported with the .params file automatically. To import the net
without the parameters, use ArrayPath set to None. Importing the net parameters can
be done with the following options: a list (ArrayList), the names (ArrayNames), or as an
association (ArrayAssociation). This is shown in Figure 9-45.

```
In[79]:=Row[Dataset[Import[FileNameJoin[{FileDirectory,"MxNet.
json"}],{"MXNet",#}]]&/@{"ArrayAssociation","ArrayList","ArrayNames"}]
Out[79]=
```

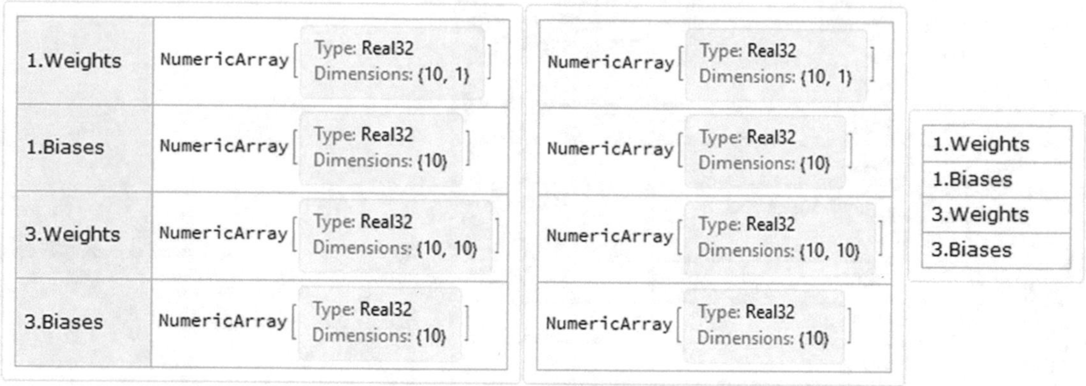
1.Weights
NumericArray[Type: Real32 Dimensions: {10, 1}]
1.Biases
NumericArray[Type: Real32 Dimensions: {10}]
3.Weights
NumericArray[Type: Real32 Dimensions: {10, 10}]
3.Biases
NumericArray[Type: Real32 Dimensions: {10}]
NumericArray[Type: Real32 Dimensions: {10, 1}]
NumericArray[Type: Real32 Dimensions: {10}]
NumericArray[Type: Real32 Dimensions: {10, 10}]
NumericArray[Type: Real32 Dimensions: {10}]
1.Weights
1.Biases
3.Weights
3.Biases

Figure 9-45. *Different import options of the MXNet format*

The elements of the net to import are InputNames, LayerAssociation, Net (import the network as a NetGraph or NetChain), NodeDataset (a dataset of the nodes of the MXNet), NodeGraph (nodes graph of the MXNet), NodeGraphPlot (plot of nodes of the MXNet), and UninitializedNet (the same as ArrayPath → None). The next dataset shows a few of the options listed before Figure 9-46.

```
In[80]= {Import[FileNameJoin[{FileDirectory,"MxNet.json"}],{"MXNet",
"NodeDataset"}],Import[FileNameJoin[{FileDirectory,"MxNet.json"}],
{"MXNet","NodeGraphPlot"}]}//Row
Out[80]=
```

	op	attrs	inputs
Input	null		
1.Weights	null		
1.Biases	null		
1	FullyConnected	2 total ›	{0, 0, 0}
		3 total ›	
2$0	relu		{3, 0, 0}
3.Weights	null		
3.Biases	null		
3	FullyConnected	2 total ›	{4, 0, 0}
		3 total ›	
4$0	softmax	1 total ›	{7, 0, 0}
5	ElementWiseSum	1 total ›	{3, 0, 0}
		3 total ›	
6$0	_Mul		{3, 0, 0}
		2 total ›	
Output1	identity		{8, 0, 0}
Output2	identity		{10, 0, 0}

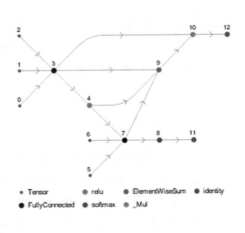

Figure 9-46. *Node dataset and MXNet ops plot*

Some operations between the Wolfram Language and MXNet are not reversible. If you pay attention, the network input, exported to MXNet format, was set as a real number, unlike the network input imported in MXNet format, which marks that the input is an array without specifying dimensions.

When constructing a neural network, there is no restriction on how many net chains or net graphs a net can have. For instance, the next example is a neural network from the Wolfram Neural Net Repository, which has a deeper sense of construction (Figure 9-47). This net is called CapsNet, which is used to estimate the depth map of an image. To consult the net, enter NetModel["CapsNet Trained on MNIST Data", "DocumentationLink"] for the documentation webpage; for notebook on the Wolfram Cloud enter NetModel["CapsNet Trained on MNIST Data", "ExampleNotebookObject"] or just ExampleNotebook for the desktop version.

```
In[81]:= NetModel["CapsNet Trained on MNIST Data"]
Out[81]=
```

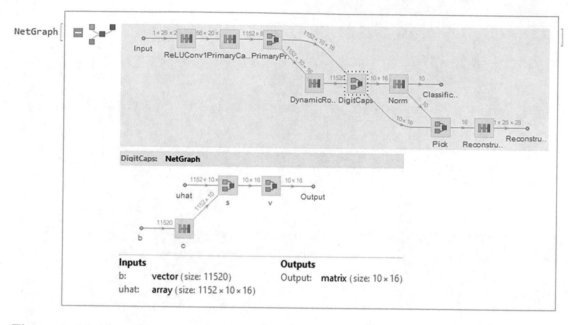

Figure 9-47. *CapsNet neural net model*

CHAPTER 10

Neural Network Framework

In this section, we will see how to train a neural network model in the Wolfram Language, how to access the results, and the trained network. We will review the basic commands to export and import a net model. We end the chapter with the exploration of the Wolfram Neural Net Repository and the review of the LeNet network model.

Training a Neural Network

The Wolfram Language contains a very useful command that automates the process of training a neural network model. This command is NetTrain. Training a neural network consists of fine-tuning the internal parameters of the neural network. The whole point of this is that the parameters can be learned during the process of training. This general process is done by an optimization algorithm called gradient descent. This, in turn, is computed with the backpropagation algorithm.

Data Input

With NetTrain, data can be entered in different forms. First the net model goes as the first argument, followed by the input → target, {inputs, ...} → {target, ...} or the name of the data or dataset. Once the net model is defined, the next argument is the data, followed by an optional argument of All. The option All will create a NetTrainResultsObject, which is used to show the NetTrain results panel after the computation and to store all relevant information about the trained model. The options for training the model are entered as last arguments. Common options used in layers and containers are available in NetTrain.

J. Villalobos Alva, *Beginning Mathematica and Wolfram for Data Science*,
https://doi.org/10.1007/978-1-4842-6594-9_10

In the next example, we will use the perceptron model to build a linear classifier. The data to be classified is shown in the next plot (Figure 10-1).

```
In[1]:=
Plt=ListPlot[{{{-1.8,-1.5},{-1,-1.7},{-1.5,-1},{-1,-1},{-0.5,-1.2},
{-1,-0.7}},{{1,1},{1.7,1},{0.5,2},{0.1,0.3},{0.5,1},{0.6,1.3}}},
PlotMarkers→"OpenMarkers",Frame→True,PlotStyle→{Green,Red}]
Out[1]=
```

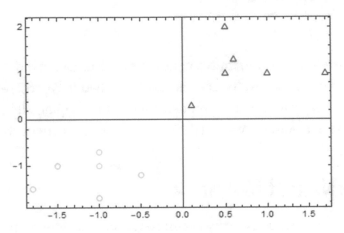

Figure 10-1. *ListPlot showing two different plot points*

Let us define the data, target values, and the training data.

```
In[2]
Data={{-1.8,-1.5},{-1,-1.7},{-1.5,-1},{-1,-1},{-0.5,-1.2},{-1,-0.7},{1,1},
{1.7,1},{0.5,2},{0.1,0.3},{0.5,1},{0.6,1.3}};
Target={-1,-1,-1,-1,-1,-1,1,1,1,1,1,1};
TrainData=MapThread[#1→ {#2}&,{Standardize[Data],Target},1];
```

Next let's define the net model.

```
In[3]:= Model=NetChain[{LinearLayer[1,"Input"→ 2],ElementwiseLayer[Ra
mp[#]&]}];
```

Training Phase

Having the data prepared and the model, we proceed to train the model. Once the training begins, a progress information panel appears, with four main results.

1. Summary: contains relevant information about the batches, rounds, and time rates

2. Data: involves processed data information

3. Method: shows the method used, batch size, and device used for training

4. Round: the current state of loss value

```
In[4]:= Net=NetTrain[Model,TrainData,All,LearningRate→0.01,PerformanceGoal
→"TrainingSpeed",TrainingProgressReporting→"Panel",TargetDevice→"CPU",
RandomSeeding→88888,WorkingPrecision→"Real64"]
Out[4]=
```

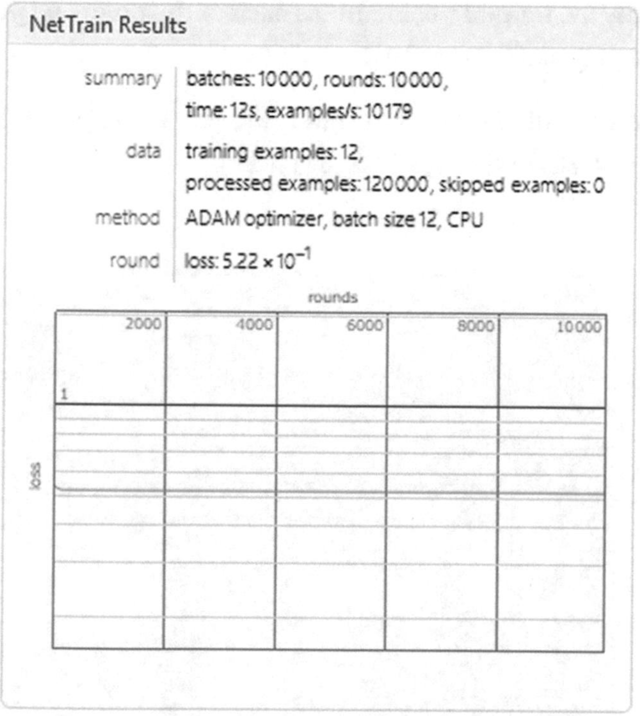

Figure 10-2. *NetTrainResultsObject*

Figure 10-2 shows the plot of the loss against the training rounds. The Adam optimizer is a variant of the Stochastic gradient descent that is seen later on. The object generated is called NetTrainResultsObject (Figure 10-2).

Model Implementation

Once the training is done, getting the trained net and model implementation is as follows in Figure 10-3.

```
In[5]:= TrainedNet1=Net["TrainedNet"]
Out[5]=
```

Figure 10-3. *Extracted trained net*

Now let's see how the trained net identifies each of the points by plotting the boundaries with density plot (Figure 10-4).

```
In[6]:= Show[DensityPlot[TrainedNet1[{x,y}],{x,-2,2},{y,-3,3},PlotPoints→5
0,ColorFunction→(RGBColor[1-#,2*#,1]&)],Plt]
Out[6]=
```

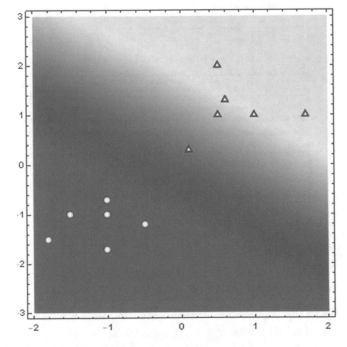

Figure 10-4. *Net classification plot*

Seeing the graphic, we can see that the boundaries are not well defined and that points near zero might be misclassified. This can be attributed because the ramp function gives 0 if it receives any negative number, but for any positive value, it returns that value. We can see that still this model can be improved perhaps by changing the activation function to a hyperbolic tangent to have robust boundaries.

Batch Size and Rounds

In the event that the batch size is not indicated, it will have an automatic value, almost always a value of 64, or powers of two. Remember that the batch size indicates the number of examples that the model uses in training before updating the internal parameters of the model. The number of batches is the division of the examples within the training dataset by the size of the batch. The processed examples are the number of rounds (epochs) multiplied by the number of training examples. In general, the batch size is chosen so that it evenly divides the size of the training set.

The option MaxTrainingRounds determines the number of times the training dataset is passed through, during the training phase. When you go through the entire training set just once, it is known as an epoch. To better understand this, in the earlier example,,. a batch size of 12 was automatically chosen, which is equal to the number of examples in the training set. This means that for epoch or round, enters a batch of, 12/12 -> 1. Now the number of epochs was automatically chosen to 10000, this tells us that there will be 1 * 10000 batches. Also, the number of processed examples will be 12 * (10000) which is equal to 120000. In case the batch size does not evenly divide the training set, it means that the final batch will have fewer examples than the other batches. Furthermore, adding a loss function layer to the container or adding the loss with the option LossFunction -> "Loss layer" has the same effect. In this case, we will use the MeanSquaredLossLayer as the loss function option and change the activation function to Tanh[x]. And setting the Batchsize -> 5 and MaxTrainingRounds -> 1000.

```
In[7]:= Net2=NetTrain[NetChain[{
LinearLayer[1,"Input"-> 2],ElementwiseLayer[Tanh[#]&]}],TrainData,All,Lear
ningRate->0.01,PerformanceGoal->"TrainingSpeed",TrainingProgressReporting-
>"Panel",TargetDevice->"CPU",RandomSeeding->88888,
WorkingPrecision->"Real64",LossFunction->MeanSquaredLossLayer[],BatchSize->5,
MaxTrainingRounds->1000]
Out[7]=
```

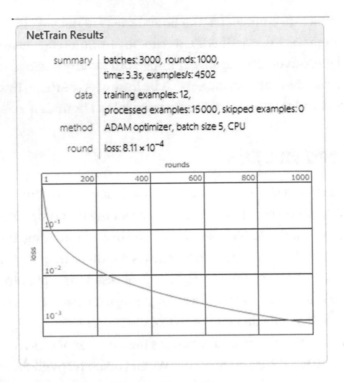

Figure 10-5. *Training results of the Net2*

We see that the loss has dropped considerably (Figure 10-5). Let us see how the classification is.

```
In[8]:= TrainedNet2=Net2["TrainedNet"];
Show[DensityPlot[TrainedNet2[{x,y}],{x,-2,2},{y,-3,3},PlotPoints-
>50,ColorFunction->(RGBColor[1-#,2*#,1]&)],Plt]
Out[8]=
```

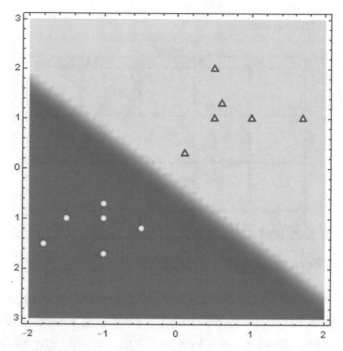

Figure 10-6. *Net2 classification plot*

We can see how the two boundaries are better denoted (Figure 10-6). The previous models represent a prediction of a linear layer, in which this classification is compared with the targets so that the error is less and less.

To obtain the graph that shows the value of the error according to the number of rounds that are carried out in the training, we do it through the properties of the trained network. We can also see how the network model looks once the loss function is added.

```
In[9]:= Dataset[{Association["LossPlot"-> Net2["LossPlot"]],Association
["NetGraph"-> Net2["TrainingNet"]]}]
Out[9]=
```

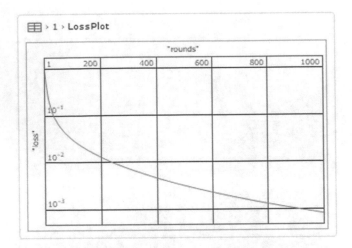

Figure 10-7. *LossPlot contained in the dataset*

In Figure 10-7, we see the graph of the loss as it decreases rapidly according to the number of rounds. To see the network used for training, execute the next code. Mathematica automatically adds a loss function to the neural network (Figure 10-8) based on the layers of the model.

```
In[10]:= Net2["TrainingNet"]
Out[10]=
```

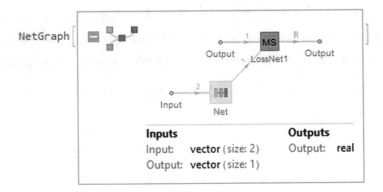

Figure 10-8. *Network model before the training phase*

To see all the properties of the model, we add the string Properties as an argument.

```
In[11]:= Net2["Properties"]
Out[11]= {ArraysLearningRateMultipliers,BatchesPerRound,BatchLossList,Batch
Measurements,BatchMeasurementsLists,BatchSize,BestValidationRound,Checkpoint
ingFiles,ExamplesProcessed,FinalLearningRate,FinalPlots,InitialLearningRate,
InternalVersionNumber,LossPlot,MeanBatchesPerSecond,MeanExamplesPerSecond,
NetTrainInputForm,OptimizationMethod,ReasonTrainingStopped,RoundLoss,Round
LossList,RoundMeasurements,RoundMeasurementsLists,RoundPositions,Skipped
TrainingData,TargetDevice,TotalBatches,TotalRounds,TotalTrainingTime,TrainedNet,
TrainingExamples,TrainingNet,TrainingUpdateSchedule,ValidationExamples,
ValidationLoss,ValidationLossList,ValidationMeasurements,ValidationMeasure
mentsLists,ValidationPositions}
```

Training Method

Let us see the training method for the previous network with OptimizationMethod. There are variants of the gradient descent algorithm, which are related to the term batch size. The first one is the stochastic gradient descent (SGD). The SGD takes a single training batch at a time before taking another step. This algorithm goes through the training examples in a stochastic form—that is, without a sequential pattern, and only one example at a time.

The second variant is the batch gradient descent, meaning that the batch size is set to the size of the training set. This method utilizes all of the training examples and makes only one update of the internal parameters.

And the third variant is the mini-batch gradient descent, which consists of dividing the training set into partitions lesser than the whole dataset in order to frequently update the internal parameters of the model, to achieve convergence. To see for a mathematical of the SGD and mini-batch SGD, visit the article, "Efficient Mini-Batch Training for Stochastic Optimization," by Mu Li, Tong Zhang, Yuqiang Chen, and Alexander J. Smola (2014, August: pp. 661-670; In *Proceedings of the 20th ACM SIGKDD international conference on Knowledge discovery and data mining*).

```
In[12]:= Net2["OptimizationMethod"]
Out[12]= {ADAM,Beta1->0.9,Beta2->0.999,Epsilon->1/100000,GradientClipping-
>None,L2Regularization->None,LearningRate->0.01,LearningRateSchedule-
>None,WeightClipping->None}
```

The method that automatically is chosen is the ADAM optimizer, which uses the SGD method, using a learning rate that is adapted. The other available methods are the RMSProp, SGD, and the SignSGD. Within the available methods, there are also options to indicate the learning rate, when to scale, when to use the L2 regularization, the gradient, and weight clipping.

Measuring Performance

In addition to the methods we can establish what measures to take into account during training phase. These options depend on the type of loss function that is being used and which is intrinsically related to the type of task, like classification, regression, clustering, etc. In the case of MeanSquaredLossLayer or MeanAbsoluteLossLayer, the common options are MeanDeviation, which is the absolute value of the average of the residuals. MeanSquare is the mean square of the residuals, RSquared is the coefficient of determination, and standard deviation is the root mean square of the residuals. After the training is completed the measure will appear in the net results (Figure 10-9).

```
In[13]:= Net3=NetTrain[NetChain[{LinearLayer[1,"Input"-> 2],Elementwise
Layer["SoftSign"]}],TrainData,All,LearningRate->0.01,PerformanceGoal->
"TrainingSpeed",TrainingProgressReporting->"Panel",TargetDevice-
>"CPU",RandomSeeding->88888,WorkingPrecision->"Real64",
Method->"ADAM",LossFunction->MeanSquaredLossLayer[],BatchSize-
>5,MaxTrainingRounds->1000,TrainingProgressMeasurements-
>{"MeanDeviation","MeanSquare","RSquared","StandardDeviation"} ]
Out[13]=
```

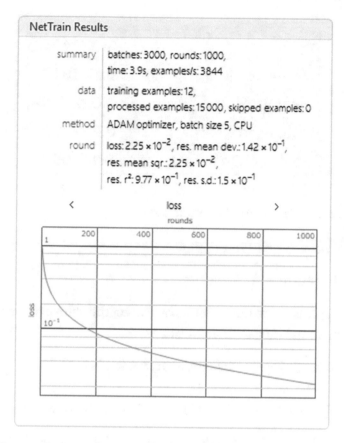

Figure 10-9. *Net results with new measures added*

Model Assessment

To access the values of the measures chosen, use the NetResultsObject. In the case of the training set values, these are found in the properties of RoundLoss (gives the average value of the loss), RoundLossList (returns the average values of the loss during training), RoundMeasurements (the measurements of the training of the last round), and RoundMeasurementsLists (the specified measurements for each round). This is depicted in Figure 10-10.

```
In[14]:= Net3[#]&/@{"RoundMeasurements"}//Dataset[#]&
Out[14]=
```

Loss	0.0224962
MeanDeviation	0.141848
MeanSquare	0.0224962
RSquared	0.977403
StandardDeviation	0.149987

Figure 10-10. *Dataset with the new measures*

To get all the plots use FinalPlots option.

```
In[15]:= Net3["FinalPlots"]//Dataset;
```

To replicate the plots of the measurements, extract the values of the measurements of each round with RoundMeasurementsLists.

```
In[16]:=Measures=Net3[#]&/@{"RoundMeasurementsLists"};
Keys[Measures]
Out[16]= {{Loss,MeanDeviation,MeanSquare,RSquared,StandardDeviation}}
```

Let us plot the values for each round, starting with Loss and finishing with StandardDeviation. We can also see how the network model makes the classification boundaries (Figure 10-11).

```
In[17]:=
TrainedNet3=Net3["TrainedNet"];
Grid[{{ListLinePlot[{Measures[[1,1]](*Loss*),Measures[[1,2]]
(*MeanDeviation*),Measures[[1,3]](*MeanSquare*),Measures[[1,4]]
(*RSquared*),Measures[[1,5]](*StandardDeviation*)},PlotStyle->Tab
le[ColorData[101,i],{i,1,5}],Frame->True,FrameLabel->{"Number of
Rounds",None},PlotLabel->"Measurements Plot",GridLines->All,
wPlotLegends->SwatchLegend[{Style["Loss",#],Style["MD",#],Style["MS",#],
Style["RS",#],Style["STD",#]},LegendLabel->Style["Measurements",#],
LegendFunction->(Framed[#,RoundingRadius->5,Background->LightGray]&)],
ImageSize->Medium]&[Black],Show[DensityPlot[TrainedNet3[{x,y}],{x,-2,2},{y,-3,3},
PlotPoints->50,ColorFunction->(RGBColor[1-#,2*#,1]&)],Plt,ImageSize-> 200]}}]
Out[17]=
```

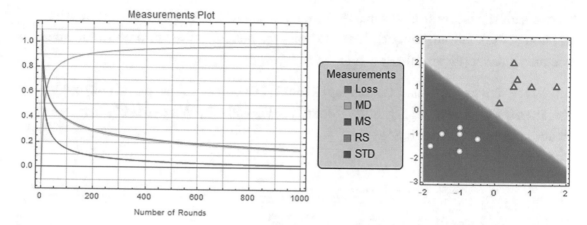

Figure 10-11. *Round measures plot and density plot*

The Loss and MeanSquared have the same values, which is why the two graphics overlap. In the case of the mean deviation and standard deviation, they have similar values but not the same. Notice that we construct three models, changing the activation function in each process. Looking at the plots, we can see how each function changes how the neural network model learns from the training data.

In the previous examples, the graphics shown were the loss plot for the training process. In the next section we will see how to plot the loss plot and the validation plot during the training phase in order to validate that a net model is actually learning during training and how well the model can perform in data never seen before (validation set).

Exporting a Neural Network

Once a net model has been trained, we can export this trained net to a WLNet format so that in the future the net can be used without the need of training. The export method also works for uninitialed network architectures.

```
In[18]:=Export["C:\\Users\\My-pc\\Desktop\\TrainedNet3.
wlnet",Net3["TrainedNet"]]
Out[18]= C:\Users\My-pc\Desktop\TrainedNEt.wlnet
```

Importing them back is done exactly as any other file, but imported elements can be specified. Net imports the net model and all initialized arrays; UninitializedNet and ArrayList imports for the numeric array's objects of the linear layers; ArrayAssociation imports for the numeric arrays in association form, and WLVersion is used to see the version of the Wolfram Language used to build the net. All of the options are shown in the next dataset (Figure 10-12).

```
In[19]:= Dataset@ AssociationMap[Import["C:\\Users\\My-pc\\Desktop\\
TrainedNet3.wlnet",#]&,{"Net","UninitializedNet","ArrayList","Array
Association","WLVersion"}]
Out[19]=
```

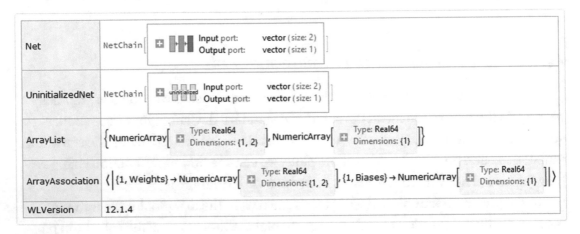

Figure 10-12. *Dataset with the available import options*

Wolfram Neural Net Repository

The Wolfram Neural Net Repository is a free access website that contains a repertoire of a variety of pre trained neural network models. The models are categorized by the type of input they receive and the type of data, be it audio, image, a numeric array, or text. Furthermore, they are also categorized by the type of task they perform, from audio analysis or regression to classification. The main page of the website is shown in Figure 10-13.

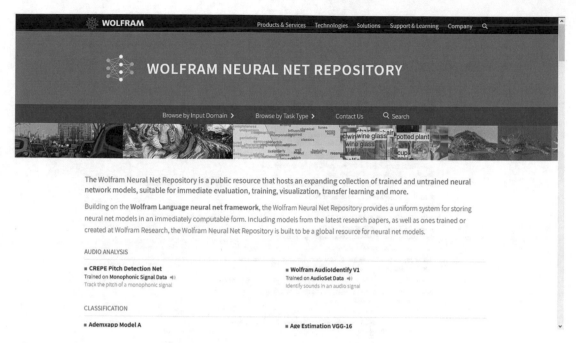

Figure 10-13. *Wolfram Neural Net Repository homepage*

To access the web page, enter the following url in your favorite browser, `https://resources.wolframcloud.com/NeuralNetRepository/` or run SystemOpen from Mathematica, which will open the web page in the system's default browser.

```
In[20]:=SystemOpen["https://resources.wolframcloud.com/
NeuralNetRepository/"];
```

Once the site is loaded, net models can be browsed by either input or by the task. The models found within this repository are built in the Wolfram Language, which allows us to make use of them within Mathematica. This leads to the models being found in a form that can be accessed either from Mathematica or from the Wolfram Cloud, for prompt execution. If we scroll down, we will see that the models are structured by name and the data used for training, along with a short description. Such is the case, for example, for the Wolfram AudioIdentify V1 network, which is trained with the AudioSet Data and identifies sounds in audio signals. To browse categories, we can choose the category from the menu. Figure 10-14 shows what the site looks like after an input category is chosen—in this case, the neural networks that receive images as inputs.

Figure 10-14. *Category based on input image*

Selecting a Neural Net Model

Once a category is chosen, it will show all of the net models associated with the selected input category. Just like with the Wolfram Data Repository, once the model is selected, it will show relevant information, like in Figure 10-15, where the selected net model is the neural network Wolfram ImageIdentify Net V1.

Figure 10-15. *Wolfram ImageIdentify Net V1*

There is the possibility of navigating from the website and downloading the notebook containing the network model, but it is also possible from Mathematica. In other words, search for network models through ResourceSearch. The example shows the search if we were interested in knowing the models of the networks that contain the word image.

```
In[21]:- ResourceSearch[{"Name"-> "Image","ResourceType"->"NeuralNet"}]//
Dataset[#,MaxItems->{4,3}]&
Out[21]=
```

Name	ResourceType	ResourceObject
Colorful Image Colorization Trained on ImageNet Competition Data	NeuralNet	ResourceObject["Colorful Image Colorization Trained on ImageNet Competition Data"]
ColorNet Image Colorization Trained on ImageNet Competition Data	NeuralNet	ResourceObject["ColorNet Image Colorization Trained on ImageNet Competition Data"]
Wolfram ImageIdentify Net V1	NeuralNet	ResourceObject["Wolfram ImageIdentify Net V1"]
Ademxapp Model A Trained on ImageNet Competition Data	NeuralNet	ResourceObject["Ademxapp Model A Trained on ImageNet Competition Data"]

rows 1–4 of 24   columns 1–3 of 6

Figure 10-16. *Resource Dataset*

The dataset that is seen in the image (Figure 10-16) has only three columns for display reasons, but using the slider you can navigate through the entire dataset. The columns not shown in the image are Description, Location, and DocumentationLink. The last column provides the link that leads to the web model page.

Accessing Inside Mathematica

To access the model architecture, the object argument is added. For example, for the Wolfram ImageIdentify Net V1 Network (Figure 10-17), do the following.

```
In[22]:= ResourceSearch[{"Name"-> "Wolfram ImageIdentify","ResourceType"->
"NeuralNet"},"Object"]
Out[22]=
```

```
{ResourceObject[ ⊟ ✳   Name: Wolfram ImageIdentify Net V1 »
                       Type: NeuralNet
                       Description: Identify the main object in an image
                       ByteCount:
                       TrainingSetInformation:
                       InputDomains: Row[Image, ,]
                       TaskType: Classification
                       Keywords: ImageIdentify, object classification
                       Data Location: Local Cloud
                       UUID: 044dd5d5-5895-4252-889f-f67729b1a6d3
                       Version: 1.10.0
                       Elements: ConstructionNotebook, EvaluationNet, UninitializedEvaluationNet, ConstructionNotebookExpression, EvaluationExample  ]}
```

Figure 10-17. *Wolfram ImageIdentify Net V1 resource*

Note   To make sure there is no problem accessing the Wolfram Net Repository from Mathematica, make sure you are logged in into the Wolfram Cloud or your Wolfram account.

Accessing the pretrained model. The next code is suppressed here, but removing the semicolon returns the NetChain object of the pretrained neural network.

```
In[23]:=ResourceSearch[{"Name"-> "Wolfram ImageIdentify","ResourceType"->
"NeuralNet"},"Object"][[1]]//ResourceData;
```

Retrieving Relevant Information

It is worth mentioning that information about the model is accessed from the ResourceObject. Following is the relevant information from the ImageIdentify model in the form of a dataset (Figure 10-18). To see all information in the dataset format, type
`ResourceObject ["Wolfram ImageIdentify Net V1"][All]//Dataset [#] &.`

```
In[24]:=Dataset[AssociationMap[ResourceObject["Wolfram ImageIdentify Net
V1"][#]&,Map[ToString,{Name,RepositoryLocation,ResourceType,ContentElements,
Version,Description,TrainingSetData,TrainingSetInformation,InputDomains,
TaskType,Keywords,Attributes,LatestUpdate,DownloadedVersion,Format,
ContributorInformation,DOI,Originator,ReleaseDate,ShortName,Wolfram
LanguageVersionRequired},1]]]
Out[24]=
```

Name	Wolfram ImageIdentify Net V1
RepositoryLocation	https://www.wolframcloud.com/objects/resourcesyste...
ResourceType	NeuralNet
ContentElements	{ConstructionNotebook, EvaluationNet, UninitializedEvaluationNet, ConstructionNotebookExpression, EvaluationExample}
Version	1.10.0
Description	Identify the main object in an image
TrainingSetData	—
TrainingSetInformation	Internal Wolfram ImageIdentify training set, consisting of over 3 million training images and over 4,000 classes of objects (not publicly available).
InputDomains	Image
TaskType	Classification
Keywords	{ImageIdentify, object classification}
Attributes	{LocalCopyable, CloudCopyable, Multipart}
LatestUpdate	Fri 28 Feb 2020 01:00:00
DownloadedVersion	1.10.0
Format	<|EvaluationNet → WLNet, UninitializedEvaluationNet → WLNet, ConstructionNotebookExpression → NB, EvaluationExample → WXF|>
ContributorInformation	<|PublisherID → Wolfram, DisplayName → Wolfram Research|>
DOI	https://doi.org/10.24097/wolfram.34204.data
Originator	Wolfram Research
ReleaseDate	Mon 20 Feb 2017 17:00:00
ShortName	Wolfram-ImageIdentify-Net-V1

rows 1–20 of 21

Figure 10-18.   Dataset of some properties of the Wolfram ImageIdentify Net V1

Here, in a few steps, is the way to access the trained neural network in addition to a lot of relevant information associated with the neural network. It should be noted that the process is also used to find other resources that are in the Wolfram Cloud or local resources, not only neural networks, since in general ResourceSearch looks for an object that is within the Wolfram Resource System. Such is the case of the neural network models that are in the Wolfram Neural Net Repository.

LeNet Neural Network

Now, in the following example, we are going to see a neural network model with the name of LeNet. Despite being able to access the model from a Wolfram resource as we saw previously, it is possible to perform operations with networks found in the Wolfram Neural Net Repository with the NetModel command. To get a better idea of how this network is used, let's first look at the description of the network, its name, how it is used, and where it was proposed for the first time.

LeNet Model

The neural network LeNet is a convolutional neuronal network that is within the field of deep learning. The neural network LeNet is recognized as one of the first convolutional networks that promoted the use of deep learning. This network was used for character recognition, to identify handwritten digits. Today there are different architectures based on LeNet neural network architecture, but we will focus on the version found in the Wolfram Neural Net Repository. This architecture consists of four key operations: convolution, non-linearity, subsampling, or pooling and classification. To learn more about the LeNet convolutional neural network, see *Neural Networks and Deep Learning: A Textbook* by Charu C. Aggarwal, (2018: Springer).

With NetModel we can obtain information about the LeNet network that has been previously trained.

```
In[25]:= NetModel["LeNet Trained on MNIST Data",#]&/@{"Details","ShortName"
,"TaskType","SourceMetadata"}//Column
Out[25]=
This pioneer work for image classification with convolutional neural nets
was released in 1998. It was developed by Yann LeCun and his collaborators
at AT&T Labs while they experimented with a large range of machine learning
solutions for classification on the MNIST dataset.
LeNet-Trained-on-MNIST-Data
{Classification}
<|Citation->Y. LeCun, L. Bottou, Y. Bengio, P. Haffner, "Gradient-Based
Learning Applied to Document Recognition," Proceedings of the IEEE,
86(11), 2278-2324 (1998),Source->http://yann.lecun.com/exdb/lenet,Date-
>DateObject[{1998},Year,Gregorian,-5.]|>
```

Note   To access all the properties of a model with NetModel, add Properties as the second argument. NetModel ["LeNet Trained on MNIST Data", "Properties"].

The input that this model receives consists of images in the grayscale with a size of 28 x 28, and the performance of the model is 98.5%.

```
In[26]:= NetModel["LeNet Trained on MNIST Data",#]&/@{"TrainingSetInformati
on","InputDomains","Performance"}//Column
Out[26]= MNIST Database of Handwritten Digits, consisting of 60,000
training and 10,000 test grayscale images of size 28x28.
{Image}
This model achieves 98.5% accuracy on the MNIST dataset.
```

MNIST Dataset

This network is used for rating, just as it appears in TaskType. The digits are in a database known as the MNIST database. The MNIST database is an extensive database of handwritten digits (Figure 10-19) that contains 60,000 images for training and 10,000 images for testing, the latter being used to get a final estimate of how well the neural net model works. To observe the complete dataset, we load it from the Wolfram Data Repository with ResourceData and with ImageDimensions verify that the dimensions of the pictures are 28 x 28 pixels.

```
In[27]:= (*This is for seven elements randomly sampled,but you can check
the whole data set.*)
TableForm[
SeedRandom[900];
RandomSample[ResourceData["MNIST","TrainingData"],7],TableDirections->Row]
Map[ImageDimensions,%[[1;;7,1]]](*Test set : ResourceData["MNIST","TestData"] *)
Out[27]//TableForm=

Out[28]=
{{28,28},{28,28},{28,28},{28,28},{28,28},{28,28},{28,28}}
```

Figure 10-19. *Random sample of the MNIST training set*

Figure 10-19 shows the images of the digits and the class to which they apply as well as the dimensions of each image. We extract the sets, training set, and the test set, which we will use later.

```
In[29]:={TrainData,TestData}={ResourceData["MNIST","TrainingData"], Resource
Data["MNIST","TestData"] };
```

LeNet Architecture

Let us start by downloading the neural network from the NetModel command, which extracts the model from the Wolfram Neural Net Repository. In the next exercise, we will load the network that has not been trained since we will do the training and validation process. It should be noted that the LeNet model in the Wolfram Language is a variation of the original architecture (Figure 10-20).

```
In[30]:= UninitLeNet=NetModel["LeNet Trained on MNIST Data","Uni
nitializedEvaluationNet"](*To work locally with the untrained model:
NetModel["LeNet"]*)
Out[30]=
```

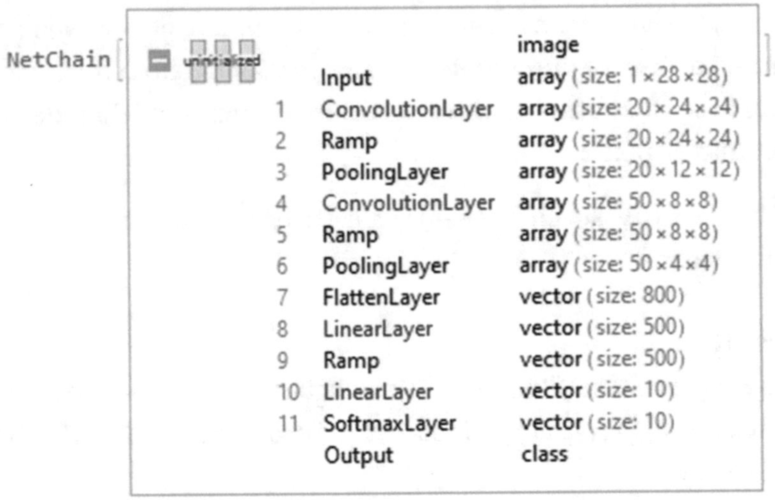

		image
NetChain	Input	array (size: 1 × 28 × 28)
1	ConvolutionLayer	array (size: 20 × 24 × 24)
2	Ramp	array (size: 20 × 24 × 24)
3	PoolingLayer	array (size: 20 × 12 × 12)
4	ConvolutionLayer	array (size: 50 × 8 × 8)
5	Ramp	array (size: 50 × 8 × 8)
6	PoolingLayer	array (size: 50 × 4 × 4)
7	FlattenLayer	vector (size: 800)
8	LinearLayer	vector (size: 500)
9	Ramp	vector (size: 500)
10	LinearLayer	vector (size: 10)
11	SoftmaxLayer	vector (size: 10)
	Output	class

Figure 10-20. *LeNet architecture*

We see that the LeNet network in the Wolfram Neural Net Repository is built from 11 layers. The layers that appear in red are layers with learnable parameters: two convolutional layers and two linear layers.

MXNet Framework

With the use of the MXNet framework, let us first visualize the process of this network through the MXNet operation graph (Figure 10-21).

```
In[31]:=Information[UninitLeNet,"MXNetNodeGraphPlot"]
Ouy[31]=
```

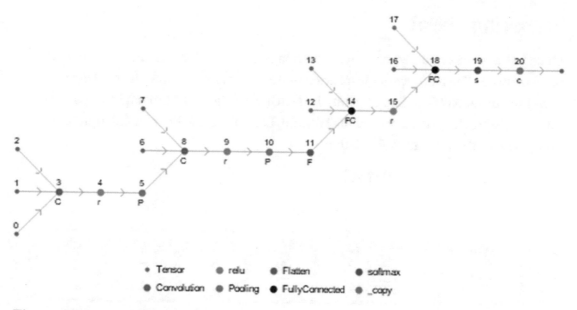

Figure 10-21. *MXNet graph*

LeNet architecture starts at the input with the encoder that converts the image to a numeric array, followed by the first operation that is a convolution that returns a 20-feature map, with a rectified linear unit activation function in nodes 3 and 4. Then the first max-pooling operation (subsampling layers) that selects the maximum value in the pooling node 5. Then it comes to the second convolutional operation, which returns a 50-feature map also with a rectified linear unit activation function in nodes 8 and 9. The last convolution operation is followed by another max-pooling operation (node 10), followed by flattening operation (node 11), which flattens the output of the pooling operation into a single vector. The last pooling operation gives an array of 50*4*4, and the flatten operation returns an 800-vector that is the input of the next operation. Next, we see the first fully connected layer (linear layer; node 14), the first fully connected layer has a rectified linear unit function (node 15), and the second fully connected layer, which has the softmax function (softmax layer; node 19). The last fully connected layer can be interpreted as a multilayer perceptron (MLP), which uses the softmax to normalize the output into a probability distribution to tell the probability of each class. Finally, the tensor is converted to a class with a decoder. The nodes 4, 9, 15, and 19 are the layers for non-linear operations.

Preparing LeNet

Since LeNet works as a neural network for image classification, an encoder and decoder must be used. The NetEncoder is inserted in the input NetPort, and the NetDecoder is on the output NetPort. Looking into the NetGraph (Figure 10-22) might be useful in understanding the process inside the Wolfram Language. Clicking on the input and output shows the relevant information.

```
In[32]:= NetGraph[UninitLeNet]
Out[32]=
```

Figure 10-22. *NetGraph of the LeNet model*

We can extract the encoder and decoder to inspect their infrastructure. The encoder receives an image of the dimensions of 28 x 28 of any color space and encodes the image into a color space set to grayscale, returning then an array of the size of 1 x 28 x 28. On the other hand, the decoder is a class decoder that receives a 10-size vector, which tells the probability for the class labels that are 0, 1, 2, 3, 4, 5, 6, 7, 8, and 9.

```
In[33]:={Enc=NetExtract[UninitLeNet,"Input"],Dec=NetExtract[UninitLeNet,
"Output"]}//Row;
```

Let's see first how the net model works with NetInitialize. As an example, we use an image of the number 0 in the training set.

```
In[34]:= TestNet=NetInitialize[UninitLeNet,RandomSeeding->8888];
TestNet@TrainData[[1,1]](*TrainData[[1,1]] belongs to a zero*)
Out[34]= 9
```

The net returns out that the image belongs to class 9, which means that the image is a number 9; clearly this is wrong. Let us try NetInitialize again but with the different methods available. Writing all, as the second argument to NetInitialize, overwrites any pre-existing learning parameters on the network.

```
In[35]:= {net1,net2,net3,net4}=Table[NetInitialize[UninitLeNet,All,Method-
>i,RandomSeeding->8888],{i,{"Kaiming","Xavier","Orthogonal","Identity"}}];
{net1[TrainData[[1,1]]],net2[TrainData[[1,1]]],net3[TrainData[[1,1]]],net4[
TrainData[[1,1]]]}
Out[35]= {9,9,5,3}
```

Every net model fails to classify the image in the correct class. This is because
the neural network has not been trained, unlike NetInitialize, which only randomly
initializes the learnable parameters but without performing proper training. This is why
NetInitialize fails to correctly classify the image given. But first, we are going to establish
the network graph to better illustrate the idea (Figure 10-23).

```
In[36]:= LeNet=NetInitialize[NetGraph[<|"LeNet NN"->UninitLeNet,"LeNet
Loss"->CrossEntropyLossLayer@"Index"|>,{NetPort@"Input"->"LeNet NN","LeNet
NN"->NetPort@{"LeNet Loss","Input"},NetPort@"Target"->NetPort@{"LeNet Loss"
,"Target"}}],RandomSeeding->8888]
Out[36]=
```

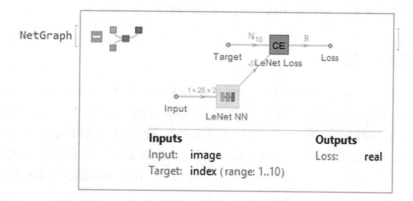

Figure 10-23. *LeNet ready for training*

Before we proceed to train the net, we need to make the validation set suited for the
CrossEntropyLossLayer in the target input, because the classes start at 0 and end in 9, and
the Index target starts at 1 and goes on. So, the target input needs to be between 1 and 10.

```
In[37]:=
]TrainDts=Dataset@Join[AssociationThread["Input"->#]& /@Keys[TrainData],
AssociationThread["Target"-> #]&/@Values[TrainData]+1,2];
TestDts=Dataset@Join[AssociationThread["Input"->#]& /@Keys[TestData],
AssociationThread["Target"-> #]&/@Values[TestData]+1,2];
```

The training set and validation set have the form of a dataset. Only four random samples are shown in Figure 10-24.

```
In[38]:=BlockRandom[SeedRandom[999];
{RandomSample[TrainDts[[All]],4],RandomSample[TestDts[[All]],4]}]
Out[38]=
```

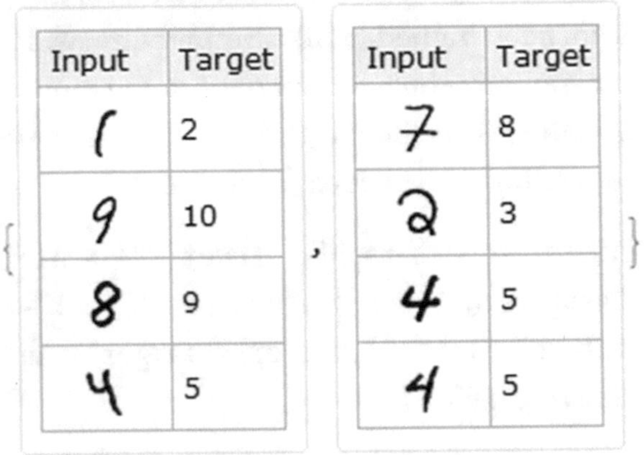

Input	Target
(	2
9	10
8	9
4	5

Input	Target
7	8
2	3
4	5
4	5

Figure 10-24. *Dataset of the training and test set*

LeNet Training

Now that we grasp the process of this neural net model, we can start now proceeding to train the neural net model. With NetTrain we gradually modify the learnable parameters of the neural network to reduce the loss. The next training code is set with the options seen in the previous section, but here we add new options that are also available for training. The first one is TrainingProgressMeasurements. With TrainingProgressMeasurements, we can specify that measures such as accuracy, precision, and so on are measured during the training phase either by round or batch. The ClassAveraging is used to specify to get the macro-average or the micro-average of the measurement specified <|"Measurement" -> "measurement" (Accuracy, RSquared, Recall, MeanSquared, etc.),"ClassAveraging"->"Macro"|>.

The second option is TrainingStoppingCriterion, which is used to add an early stopping to avoid overfitting during the training phase, based on different criteria, such as stopping the training when the validation loss is not improving, measuring the absolute or relative change of a measurement (accuracy, precision, loss, etc.), or stopping the training when the loss or other criteria does not improve after a certain number of rounds <|Criterion->"measurement" (Accuracy, Loss, Recall, etc.),"Patience"-> # of rounds|>.

```
In[39]:= NetResults=NetTrain[LeNet,TrainDts,All,ValidationS
et->TestDts,MaxTrainingRounds->15,BatchSize->2096,LearningRate-
>Automatic,Method->"ADAM",TargetDevice->"CPU",PerformanceGoal-
>"TrainingMemory",WorkingPrecision->"Real32",RandomSeeding->99999,Train
ingProgressMeasurements->{<|"Measurement"->"Accuracy","ClassAveraging"-
>"Macro"|>, <|"Measurement"->"Precision","ClassAveraging"-
>"Macro"|>,<|"Measurement"->"F1Score","ClassAveraging"-
>"Macro"|>,<|"Measurement"->"Recall","ClassAveraging"-
>"Macro"|>,<|"Measurement"->"ROCCurvePlot","ClassAveraging"-
>"Macro"|>,<|"Measurement"->"ConfusionMatrixPlot","ClassAveragi
ng"->"Macro"|> },TrainingStoppingCriterion-> <|"Criterion"->"Loss","Absolut
eChange"->0.001|>]
Out[39]=
```

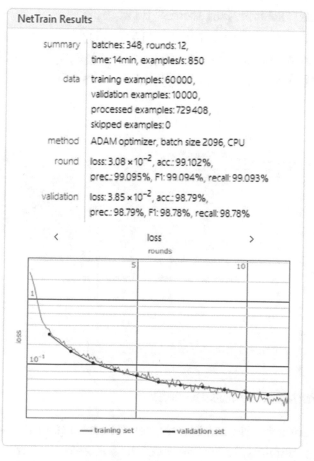

Figure 10-25. *Net results of LeNet training*

The final results of the training phase are depicted in the Figure 10-25. Extracting the trained model and appending the net encoder and decoder is done because the trained net does not come with an encoder and decoder at the input and output ports.

```
In[40]:=NetExtract[NetResults["TrainedNet"],"LeNet NN"];
TrainedLeNet=NetReplacePart[%,{"Input"->Enc,"Output"->Dec}];
```

LeNet Model Assessment

The next grid (Figure 10-26) shows the tracked measurements and plots of the training set. The measurements of the training set are in the RoundMeasurements property. To get the list of the values in each round, use RoundMeasurementsLists. The performance of the training set is assessed with the round measurements and the test set is assessed with the validation measurements. Also, in both cases the ROC curves and the confusion matrix plot are shown.

```
In[41]:=NetResults["RoundMeasurements"][[1;;5]];
Normal[NetResults["RoundMeasurements"][[6;;7]]];
Grid[{{Style["RoundMeasurements",#1,#2],Style[%[[1,1]],#1,#2],Style[%[[2,1]],
#1,#2]},{Dataset[%%],%[[1,2]],%[[2,2]]}},Dividers->Center]&[Bold,FontFamily
->"Alegreya SC"]
Out[41]=
```

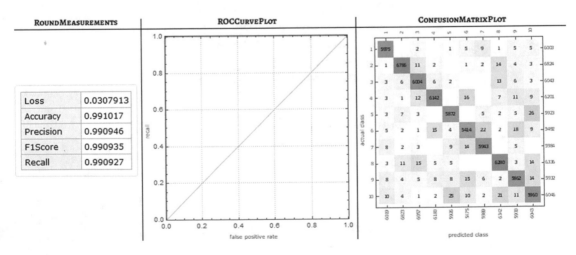

Figure 10-26. *Training set measurements*

To see how the model performed on the validation set (Figure 10-27), see ValidationMeasurements. To get the list of the values in each round, use ValidationMeasurementsLists.

```
In[42]:=NetResults["ValidationMeasurements"][[1;;5]];
Normal[NetResults["ValidationMeasurements"][[6;;7]]];
Grid[{{Style["ValidationMeasurements",#1,#2],Style[%[[1,1]],#1,#2],Style[%[
[2,1]],#1,#2]},{Dataset[%%],%[[1,2]],%[[2,2]]}},Dividers->Center]&[Bold,Fo-
ntFamily->"Alegreya SC"]
Out[42]=
```

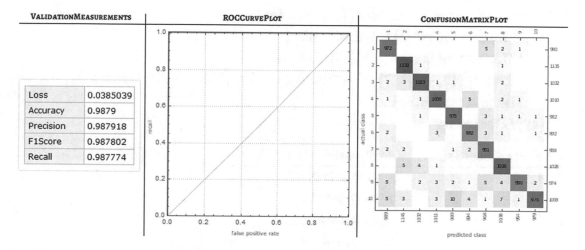

Figure 10-27. *Test set measurements*

Testing LeNet

Having finished the training and reviewed the round measures and validation measures,
we are now erady to test the trained LeNet neural network with some difficult images to
see how it performs (Figure 10-28).

```
In[43]:=
Expls=Keys[{TestData[[2150]],TestData[[3910]],TestData[[6115]],TestData[[60
11]],TestData[[7834]]}]
Out[43]=
```

{ 2, 3, 6, 6, 7 }

Figure 10-28. *Difficult examples from the MNIST test set*

The selected images belong to the numbers 2, 3, 6, 5, and 7.

```
In[44]:= TrainedLeNet[Expls,"TopProbabilities"]
Out[44]= {{2->0.998171},{3->0.999922},{6->0.587542,0->0.404588},
{6->0.971103},{7->0.99937}}
```

Write all of the results with the top probabilities with TableForm.

```
In[45]:= TableForm[Transpose@{TrainedLeNet[
Expls,{"TopDecisions",2}],TrainedLeNet[
Expls,{"TopProbabilities",2}]},TableHeadings-
>{Map[ToString,{2,3,6,5,7},1],{"Top Decisions","Top Probabilities"}},
TableAlignments->Center]
Out[45]//TableForm=
```

	Top Descisions	Top Probabilities
2	3	3 -> 0.00165948
	2	2 -> 0.998171
3	9	9 -> 0.0000534744
	3	3 -> 0.999922
6	0	0 -> 0.404588
	6	6 -> 0.587542
5	0	0 -> 0.0140468
	6	6 -> 0.971103
7	3	3 -> 0.000526736
	7	7 -> 0.99937

We can see that the trained net has misclassified the image of the number 5, because the top decisions are either a 0 or a 6, and clearly that is wrong. Also, we can see the probabilities of the top decisions. Another form to evaluate the trained net in the test set is with the use of NetMeasurements to set the net model, test set, and the interested measure. In the example, the measure of interest is the ConfusionMatrixPlot (Figure 10-29).

```
In[26]:=NetMeasurements[TrainedLeNet,TestData,"ConfusionMatrixPlot"]
Out[26]=
```

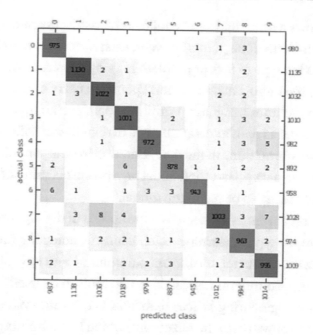

Figure 10-29. *ConfusionMatrixPlot from NetMeasurements*

Final Remarks

In summary, it can be concluded that in general terms, a road map for the general schematics, construction, testing, and implementation of a machine learning or a neural network model within the Wolfram Language scheme. This is shown in Figure 10-30.

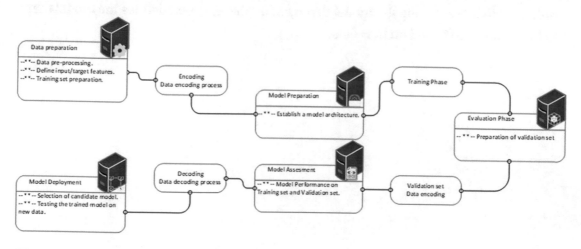

Figure 10-30. *Training set measurements*

The diagram shows a route that can be followed directly; despite this, within the route, there may be intermediate points between each process, since the route may vary depending on what type of task or problem is being solved. However, the route focuses on exposing the important and general points to carry out the construction of a model using the Wolfram Language. Within the data preparation phase, there are previous processes, such as data integration, type of data that is collected (structured or unstructured), transformations in the data, cleaning in data modules, and so on. So before moving on to the next phase, there must be a pre-processing of the data, with the intention to have data ready to be fed to the model.

Model preparation covers aspects such as the choice of the algorithm or the methods to use, depending on the type of learning; establishing or detecting the structure of the model; and defining the characteristics, input parameters, and type of data that will be used, whether it be text, sound, numerical data, and tools to be used. All this is linked to process called feature engineering, whose main goal is to extract valuable attributes from data. This is needed to move on to the next point, which is the training phase.

The evaluation phase and model assessment consist of defining the evaluation metrics, which vary according to the type of task or problem that is being solved in addition to preparing the validation that will be used later. At this point it is necessary to emphasize that the preparation of the model, training, and evaluation and assessment can be an iterative process, which can include tuning of hyperparameters, adjustments on algorithm techniques, and model configurations such as internal model features. The purpose is to establish the best possible model that is capable of delivering adequate results and finally reaching the model deployment phase, which defines the model that will be chosen and tested on new data.

APPENDIX A

Installing Mathematica

Wolfram Technologies has a free 15-day trial with the complete platform, by creating a Wolfram ID account with some basic information like name, country, address, etc. In order to get this, visit the `https://www.wolfram.com/mathematica/trial/`.

Having downloaded the execution file (the type of file depends on the operating system you have), start the execution file and proceed with installing it. Once started, the download manager appears and starts downloading the Mathematica program. Proceeding with the setup, select the directory folder in which Mathematica should be installed. Select install all components; the other components include the WolframScript, which is the script program to run Wolfram Language in a script terminal, and can be used in, for example, a Jupyter notebook but the implementation is beyond the scope of this book. Once the setup is done, proceed to install Mathematica. Once the installation is finished, run Mathematica by clicking the desktop shortcut or look for the Wolfram Mathematica folder under the start menu of your operating system. Having opened Mathematica, it will ask you for the license key, which is provided to you via the email you used to register your Wolfram account. Having finished, you are ready to start computations inside Mathematica.

To review more in-depth installation for other environments, visit the following web page on the Wolfram reference documentation: `https://reference.wolfram.com/language/tutorial/InstallingMathematica.html`

© Jalil Villalobos Alva 2021
J. Villalobos Alva, *Beginning Mathematica and Wolfram for Data Science*,
https://doi.org/10.1007/978-1-4842-6594-9

Index

© Jalil Villalobos Alva 2021
J. Villalobos Alva, *Beginning Mathematica and Wolfram for Data Science*,
https://doi.org/10.1007/978-1-4842-6594-9

Printed in the United States
By Bookmasters